POLYNOMIAL APPROXIMATION

POLYNOMIAL APPROXIMATION

Robert P. Feinerman

Associate Professor of Mathematics
Herbert H. Lehman College
City University of New York
New York, New York

Donald J. Newman

Professor of Mathematics
Belfer Graduate School of Science
Yeshiva University
New York, New York

THE WILLIAMS & WILKINS COMPANY
BALTIMORE

COPYRIGHT ©, 1974
ROBERT P. FEINERMAN
DONALD J. NEWMAN

All rights reserved. This book is protected by copyright. No part of this book may reproduced in any form or by any means, including photocopying, or utilized by any information storage and retrieval system without written permission from the copyright owner.

Made in the United States of America

Library of Congress Catalog Card Number 73-6808
SBN 0-683-03077-9

Composed and printed at the
WAVERLY PRESS, INC.
Mt. Royal and Guilford Aves.
Baltimore, Md. 21202, U.S.A.

PREFACE

Our main purpose in writing this book has been to introduce the reader to some aspects of polynomial approximation that do not seem to be stressed in other books on the topic. The main such topic (and the underlying theme in the book) is that of the quality of approximation; that is, how well polynomials approximate. We have made no effort to make this book into an encyclopedia; rather, we have attempted to *introduce* the reader to (and hopefully interest him in) the various topics covered. Since parts (or all) of Chapters VI, VII, VIII, IX, and XI consist of fairly recent research papers, the reader is introduced to some of the work being done on the subject and could easily find research topics to pursue.

This book can be used as a textbook for a graduate or an advanced undergraduate course. In fact, courses containing most of the material in this book were taught at Yeshiva in 1964–65 (by D. J. N.), at Harvard in the fall of 1967 (by R. P. F.), at the Hebrew University in 1969–70 (by R. P. F.), and at Yeshiva in 1972–73 (by D. J. N.). The prerequisite for such a course is the usual "mathematical maturity." The student should have had the beginnings of courses in Real, Complex, and Functional Analysis. (That is, he should know what is normally covered in the first few lectures on Lebesgue integration, analytic functions, and linear functionals.) A course based on this book should cover the first five chapters and any (or all) of the last seven. The last seven chapters are independent of each other with the exception of that part of Chapter X necessary for Chapter XI.

In this book all of the functions are assumed to be real-value. That is mainly a matter of convenience. Most of the theorems apply just as well to complex-valued functions (with little or no modifications) and only a very select few (mainly in Chapter III) require either extensive modifications or are not modifiable at all.

Theorem 6.3 is the third theorem in Chapter VI. In Chapter VI itself, however, it is referred to as theorem 3. This example applies equally well to both theorems and lemmas.

CONTENTS

PREFACE	v
I. THE WEIERSTRASS THEOREM	1
1) Fourier Series	1
2) Tchebychev Polynomials	2
3) Lebesgue's Proof	4
4) Bernstein Polynomials	6
II. PRELIMINARIES	13
1) Moduli of Continuity	13
2) Compact and Pre-Compact Sets	15
3) Zeroes of Trigonometric Polynomials	17
4) Derivatives of Polynomials	18
5) Limits of Polynomials	22
III. POLYNOMIALS OF BEST APPROXIMATION	24
1) Existence	24
2) Characterization	25
3) Uniqueness	30
4) Best Trigonometric Approximation	33
IV. JACKSON'S THEOREMS	35
1) Gaussian Quadrature	35
2) Jackson's Theorem for $C[0, 1]$	36
3) Jackson's Theorem for $C^*[-\pi, \pi]$	40
4) Favard's Theorem	46
V. INVERSE THEOREMS FOR PERIODIC FUNCTIONS	49
1) Bernstein's Theorems	49
2) The Zygmund Class	53
3) Approximating Lip α	55

VI. LINEAR OPERATORS 57
  1) Summation of Fourier Series 57
  2) Bounded Linear Operators 60
  3) Korovkin's Theorem 68

VII. RATIONAL APPROXIMATION 71
  1) Rational vs. Polynomials 71
  2) Counter Examples 75
  3) Rational Approximation to $|x|$ 76

VIII. GENERALIZED POLYNOMIALS 81
  1) Best Approximation 81
  2) Lower Bounds for $E\phi(\mathcal{S})$ 83
  3) Approximation by Step Functions 89

IX. JACKSON'S THEOREM IN k-DIMENSIONS 97
  1) Lower Bounds 98
  2) Upper Bounds 101
  3) Proof of Theorem 8 102

X. COMPLETENESS 108
  1) Completeness 108
  2) Müntz's Theorem 111
  3) More Completeness Questions 115
  4) Codimension 117

XI. A MÜNTZ-JACKSON THEOREM 121
  1) The Upper Bound 124
  2) The Lower Bound 126
  3) Computing Some ϵ_A 132

XII. A UNIFIED TREATMENT OF JACKSON'S THEOREM . . . 138

BIBLIOGRAPHICAL NOTES 143

BIBLIOGRAPHY 145

INDEX OF SYMBOLS 147

INDEX 148

CHAPTER I

THE WEIERSTRASS THEOREM

The classical theorem of polynomial approximation is the famous Weierstrass theorem: Every continuous function on [0, 1] can be uniformly approximated by polynomials.

We shall prove this theorem in a number of ways and these proofs will provide an introduction to various concepts, techniques and quantities which will be used throughout this book.

(1) Fourier Series

From Fourier series, we know the following theorem: The Cesàro sum of a continuous 2π-periodic function on $[-\pi, \pi]$ converges uniformly to the function; i.e., let $f(x)$ be continuous on $[-\pi, \pi]$ and $f(-\pi) = f(\pi)$. Let the Fourier series of $f(x)$ be

$$\sum_{n=0}^{\infty} a_n \cos nx + b_n \sin nx$$

and let

$$\sigma_N(f, x) = \left(\frac{1}{N+1}\right)\bigg[(N+1)a_0 + Na_1 \cos x + Nb_1 \sin x + \cdots + a_N \cos Nx + b_N \sin Nx\bigg].$$

Then

$$\lim_{N \to \infty} |f(x) - \sigma_N(f, x)| = 0$$

uniformly for $x \in [-\pi, \pi]$.

As a corollary of this theorem, we have the following form of the Weierstrass theorem: Every continuous 2π-periodic function on $[-\pi, \pi]$ can be uniformly approximated by trigonometric polynomials.

Notation. Let $-\infty \leq a < b \leq \infty$. $C[a, b]$ is the set of all conti-

1

nuous real valued functions on $[a, b]$. $C^*[-\pi, \pi]$ is the set of all continuous, 2π-periodic, real valued functions on $[-\pi, \pi]$. Unless otherwise specified, if $f \in C[a, b]$ (or $C^*[-\pi, \pi]$), then $\|f\|$ is defined as

$$\max_{x \in [a,b]} |f(x)| \, (\text{or} \max_{x \in [-\pi,\pi]} |f(x)|).$$

Proof (1) of Weierstrass Theorem

Let $f(x) \in C[0, 1]$. Extend $f(x)$ to $[-\pi, \pi]$ such that $f(x) \in C^*[-\pi, \pi]$. We know that given any $\epsilon > 0$ there is a trigonometric polynomial $T(x)$ such that $|f(x) - T(x)| < \epsilon$ for all $x \in [-\pi, \pi]$. Since $T(x)$ is an entire function of x, its power series converges uniformly on compact subsets of the plane and in particular on $[-\pi, \pi]$. Therefore, there is a polynomial $P(x)$ such that $|T(x) - P(x)| < \epsilon$ for $x \in [-\pi, \pi]$. Therefore,

$$|f(x) - P(x)| \leq |f(x) - T(x)| + |T(x) - P(x)| < 2\epsilon$$

for $x \in [-\pi, \pi]$ and certainly for $x \in [0, 1]$. ■[1]

(2) Tchebychev Polynomials

LEMMA 1. Let $T_n(x) = \cos(n \arccos x)$ for $x \in [-1, 1]$. Then $T_n(x)$ is a polynomial of degree n.

Proof. For $n = 0$, we have $T_0(x) = \cos(0) = 1$ while for $n = 1$, we have $T_1(x) = \cos(\arccos x) = x$. If in the formula

$$\cos(n + 1)\theta + \cos(n - 1)\theta = 2\cos\theta \cos n\theta$$

we let $\theta = \arccos x$, we obtain the recursion formula

$$T_{n+1}(x) + T_{n-1}(x) = 2x \, T_n(x).$$

Combining this with our knowledge of $T_0(x)$ and $T_1(x)$, we can prove the lemma inductively. ■

$T_n(x)$ is called the Tchebychev polynomial of degree n.

LEMMA 2. Let $g(x) \in C^*[-\pi, \pi]$ be such that $g(x) = g(-x)$ for all $x \in [-\pi, \pi]$. Then $g(x)$ can be uniformly approximated by an even trigonometric polynomial (one involving only cosines).

Proof. Since $g(x)$ is even, all its sine coefficients are zero. Therefore, its Cesàro sum, which converges to it uniformly, involves only cosines. ■

Proof (2) of Weierstrass Theorem

Given $f \in C[0, 1]$, extend it evenly to $C[-1, 1]$; i. e., $f(x) = f(-x)$

---

[1] Throughout the text, ■ is used to denote the end of proofs.

for $x \in [-1, 1]$. Let $g(\theta) = f(\cos\theta)$. Then $g \in C^*[-\pi, \pi]$ and $g(\theta) = g(-\theta)$. By lemma 2, given $\epsilon > 0$ we can find $a_0, \ldots, a_N$ such that

$$\left| g(\theta) - \sum_{n=0}^{N} a_n \cos n\theta \right| < \epsilon \text{ for all } \theta \in [-\pi, \pi].$$

Now let $\theta = \arccos x$. Then $g(\theta) = f(x)$ and $\cos n\theta = T_n(x)$. We therefore have

$$\left| f(x) - \sum_{x=0}^{N} a_n T_n(x) \right| < \epsilon \text{ for all } x \in [-1, 1]$$

and certainly for all $x \in [0, 1]$. ∎

For future use, we derive here a formula for the coefficients of the Tchebychev polynomials. That is, for $k \leq n$ we will find the coefficient of x^k in $T_n(x)$. First of all we notice that, if n is even, then $T_n(x)$ has only even powers of x, whereas, if n is odd, $T_n(x)$ has only odd powers of x. (This follows trivially from $T_0(x) = 1$, $T_1(x) = x$ and $T_{n+1}(x) + T_{n-1}(x) = 2x T_n(x)$.) Thus we only have to consider the cases when k and n are either both even or both odd.

We recall the expansion of the Poisson kernel

$$\frac{1 - r^2}{1 - 2r\cos\theta + r^2} = 1 + 2 \sum_{n=1}^{N} r^n \cos n\theta$$

and let $\theta = \arccos x$. Then we get

$$\frac{1 - r^2}{1 - 2rx + r^2} = 1 + 2 \sum_{n=1}^{\infty} r^n T_n(x).$$

Thus the k-th coefficient of $T_n(x)$ is twice the coefficient of $r^n x^k$ in $(1-r^2)/(1-2rx+r^2)$. We will first find the coefficient of x^k (in terms of r) and then find the coefficient of r^n in the expansion of the coefficient of x^k. We have

$$\frac{1 - r^2}{1 - 2rx + r^2} = \frac{1 - r^2}{1 + r^2} \frac{1}{1 - (2r/1 + r^2)x}$$

$$= \left(\frac{1 - r^2}{1 + r^2}\right) \sum_{j=0}^{\infty} \left(\frac{2r}{1 + r^2}\right)^j x^j.$$

Thus the coefficient of x^k is

$$\left(\frac{1 - r^2}{1 + r^2}\right)\left(\frac{2r}{1 + r^2}\right)^k.$$

We also have

$$\left(\frac{1-r^2}{1+r^2}\right)\left(\frac{2r}{1+r^2}\right)^k = (1-r^2)(2r)^k \frac{1}{(1+r^2)^{k+1}}$$

$$= (1-r^2)(2r)^k \sum_{j=0}^{\infty} \binom{j+k}{k}(-r^2)^j$$

$$= 2^k(1-r^2) \sum_{j=0}^{\infty} (-1)^j \binom{j+k}{k} r^{2j+k}.$$

Then the coefficient of r^n is

$$2^k \left[(-1)^{\frac{n-k}{2}} \binom{\frac{n+k}{2}}{k} - (-1)^{\frac{n-k}{2}-1} \binom{\frac{n+k}{2}-1}{k} \right]$$

$$= 2^k(-1)^{\frac{n-k}{2}} \frac{2n}{n+k} \binom{\frac{n+k}{2}}{k}.$$

Thus the coefficient of x^k in $T_n(x)$ is

$$2^k(-1)^{\frac{n-k}{2}} \frac{n}{n+k} \binom{\frac{n+k}{2}}{k}$$

(if n and k are both even or both odd and otherwise the coefficient is 0). In this derivation, we have tacitly assumed that k was positive. However, it is fairly easy to show that this formula holds for $k = 0$ also.

(3) Lebesgue's Proof

We now present one more proof of Weierstrass' theorem, a proof due to Lebesgue. This one introduces us to the important role that $|x|$ plays in approximation theory. (See theorems 6.1 and 7.4, for example.) The idea of the proof is to approximate any continuous $f(x)$ by a connected line segment function (of the form $\sum_{k=0}^{N} c_k |x - k/N| + c$) and then approximate that function by polynomials.

Proof (3) of Weierstrass Theorem

Take any $f \in C[0, 1]$ and any positive integer N. We let $L(x)$ be defined by the conditions $L(k/N) = f(k/N)$ for $k = 0, \ldots, N$ and $L(x)$ is linear for $k/N \leq x \leq (k+1)/N$. Then, for $k/N \leq x \leq (k+1)/N$,

$$|f(x) - L(x)| \leq \left| f(x) - f\left(\frac{k}{N}\right) \right| + \left| f\left(\frac{k}{N}\right) - L\left(\frac{k}{N}\right) \right|$$

$$+ \left| L\left(\frac{k}{N}\right) - L(x) \right|$$

$$\le \left|f(x) - f\left(\frac{k}{N}\right)\right| + 0 + \left|L\left(\frac{k}{N}\right) - L\left(\frac{k+1}{N}\right)\right|$$

$$= \left|f(x) - f\left(\frac{k}{N}\right)\right| + \left|f\left(\frac{k}{N}\right) - f\left(\frac{k+1}{N}\right)\right|$$

which, by the uniform continuity of $f(x)$, goes to zero as N goes to infinity. Thus $f(x)$ can be uniformly approximated by $L(x)$.

This $L(x)$ can be written as $\sum_{k=0}^{N-1} c_k |x - k/N| + c$ where

$$c = \frac{1}{2}\left[f(0) + Nf\left(\frac{N-1}{N}\right) - (N-1)f(1)\right]$$

$$c_0 = \frac{N}{2}\left[-f(0) + f\left(\frac{1}{N}\right) - f\left(\frac{N-1}{N}\right) + f(1)\right]$$

$$c_k = \frac{N}{2}\left[f\left(\frac{k+1}{N}\right) - 2f\left(\frac{k}{N}\right)f + \left(\frac{k-1}{N}\right)\right]$$

$$k = 1, 2, \ldots, N-1$$

Assume that we knew that $|x|$ could be uniformly approximated on $[-1, 1]$ by polynomials. Then, for any $\lambda \in [0, 1]$, $|x - \lambda|$ could be uniformly approximated by polynomials on $[0, 1]$ (if $|P(x) - |x|| < \epsilon$ on $[-1, 1]$, $|P(x - \lambda) - |x - \lambda|| < \epsilon$ on $[0, 1]$), and then by approximating each of the terms of $L(x)$ by polynomials we would have $L(x)$ uniformly approximated by polynomials. Thus all we have to do is show that $|x|$ can be uniformly approximated by polynomials on $[-1, 1]$.

We first write

$$|x| = \sqrt{x^2} = \sqrt{1 - (1 - x^2)}.$$

For any $\rho \in (0, 1)$ and all $x \in [-1, 1]$,

$$|\sqrt{1 - (1 - x^2)} - \sqrt{1 - \rho(1 - x^2)}| \le \sqrt{1 - \rho}.$$

Thus if ρ is close to 1, $|x|$ is uniformly close to $\sqrt{1 - \rho(1 - x^2)}$. We then let $z = \rho(1 - x^2)$ and note that $\sqrt{1 - z}$ is an analytic function of z, in $|z| < 1$ (which certainly includes $\rho(1 - x^2)$ for $x \in [-1, 1]$). As such its power series converges to it uniformly in $|z| \le \rho$. Thus

$$\sqrt{1 - \rho(1 - x^2)}$$

is uniformly near something of the form

$$\sum_{n=0}^{M} a_n (\rho(1 - x^2))^n$$

for $x \in [-1, 1]$ and we have proven that $|x|$ can be uniformly approximated by polynomials on $[-1, 1]$. ∎

(4) Bernstein Polynomials

Definition. For $f(x)$ defined on $[0, 1]$, and $N = 1, 2, 3, \ldots$, let

$$B_N(f, x) = \sum_{n=0}^{N} f\left(\frac{n}{N}\right)\binom{N}{n} x^n (1-x)^{N-n}.$$

This is known as the N-th Bernstein polynomial associated with $f(x)$.

Examples:

(a) $\quad B_N(1, x) = \sum_{n=0}^{N} \binom{N}{n} x^n (1-x)^{N-n} = (x + (1-x))^N \equiv 1$

(b) $\quad B_N(x, x) = \sum_{n=1}^{N} \frac{n}{N}\binom{N}{n} x^n (1-x)^{N-n}$

$$= x \sum_{n=1}^{N} \binom{N-1}{n-1} x^{n-1} (1-x)^{N-1-(n-1)}$$

$$= x \sum_{j=0}^{N-1} \binom{N-1}{j} x^j (1-x)^{N-1-j}$$

$$= x(x + (1-x))^{N-1} = x$$

(c) $\quad B_N(x^2, x) - \dfrac{x}{N} = \sum_{n=0}^{N} \dfrac{n^2}{N^2}\binom{N}{n} x^n (1-x)^{N-n}$

$$- \sum_{n=0}^{N} \frac{n}{N^2}\binom{N}{n} x^n (1-x)^{N-n}$$

$$= \frac{N-1}{N} \sum_{n=2}^{N} \frac{n(n-1)}{N(N-1)} \binom{N}{n} x^n (1-x)^{N-n}$$

$$= \frac{N-1}{N} x^2 \sum_{n=2}^{N} \binom{N-2}{n-2} x^{n-2} (1-x)^{(N-2)-(n-2)}$$

$$= \frac{N-1}{N} x^2 \sum_{j=0}^{N-2} \binom{N-2}{j} x^j (1-x)^{N-2-j}$$

$$= \frac{N-1}{N} x^2 (x + (1-x))^{N-2} = \frac{N-1}{N} x^2.$$

Therefore

$$B_N(x^2, x) = \frac{N-1}{N} x^2 + \frac{x}{N}.$$

LEMMA 3.
$$\sum_{n=0}^{N}\left(\frac{n}{N}-x\right)^2\binom{N}{n}x^n(1-x)^{N-n} = \frac{x(1-x)}{N}.$$

Proof. The left-hand side is just
$$B_N(x^2, x) - 2xB_N(x, x) + x^2 B_N(1, x)$$
which is the same as
$$\frac{N-1}{N}x^2 + \frac{x}{N} - 2x^2 + x^2 = -\frac{x^2}{N} + \frac{x}{N} = \frac{x(1-x)}{N} \quad \blacksquare$$

THEOREM 4. (a) (Proof (4) of Weierstrass theorem). If $f(x) \in C[0, 1]$, $B_N(f, x)$ converges to $f(x)$ uniformly on $[0, 1]$.
(b) If $f(x)$ is defined and bounded on $[0, 1]$, and if $x_0 \in [0, 1]$ is a point of continuity of $f(x)$, then $B_N(f, x_0)$ converges to $f(x_0)$.

Proof. Assume $|f(x)| \leq M$ on $[0, 1]$ and take $x_0 \in [0, 1]$. Then
$$B_N(f, x_0) - f(x_0) = \sum_{n=0}^{N} f\left(\frac{n}{N}\right)\binom{N}{n}x_0^n(1-x_0)^{N-n} - f(x_0)$$
$$= \sum_{n=0}^{N}\left[f\left(\frac{n}{N}\right) - f(x_0)\right]\binom{N}{n}x_0^n(1-x_0)^{N-n}$$
$$= \sum_{n \in S_1}\left[f\left(\frac{n}{N}\right) - f(x_0)\right]\binom{N}{n}x_0^n(1-x_0)^{N-n}$$
$$+ \sum_{n \in S_2}\left[f\left(\frac{n}{N}\right) - f(x_0)\right]\binom{N}{n}x_0^n(1-x_0)^{N-n}$$

where S_1 is all n such that $|n/N - x_0| \leq 1/N^{1/4}$ and S_2 is the remaining n. We have
$$\left|\sum_{n \in S_2}\left[f\left(\frac{n}{N}\right) - f(x)\right]\binom{N}{n}x_0^n(1-x_0)^{N-n}\right| \leq 2M \sum_{n \in S_2}\binom{N}{n}x_0^n(1-x_0)^{N-n}$$
$$= 2M \sum_{n \in S_2}\left(\frac{n - Nx_0}{n - Nx_0}\right)^2\binom{N}{n}x_0^n(1-x_0)^{N-n}$$
$$\leq 2M \sum_{n \in S_2}\frac{(n - Nx_0)^2}{N^{3/2}}\binom{N}{n}x_0^n(1-x_0)^{N-n}$$
$$\leq 2M N^{1/2} \sum_{n=0}^{N}\left(\frac{n}{N} - x_0\right)^2\binom{N}{n}x_0^n(1-x_0)^{N-n}$$
$$= 2M N^{1/2}\frac{x_0(1-x_0)}{N} \leq \frac{M}{2\sqrt{N}}$$

Let
$$\epsilon_N(x_0) = \max_{n \in S_1} \left| f(x_0) - f\left(\frac{n}{N}\right) \right|.$$

Then
$$\left| \sum_{n \in S_1} \left[f\left(\frac{n}{N}\right) - f(x_0) \right] \binom{N}{n} x_0^n (1-x_0)^{N-n} \right|$$
$$\leq \epsilon_N(x_0) \sum_{n \in S_1} \binom{N}{n} x_0^n (1-x_0)^{N-n}$$
$$\leq \epsilon_N(x_0) \sum_{n=0}^{N} \binom{N}{n} x_0^n (1-x_0)^{N-n}$$
$$= \epsilon_N(x_0).$$

We therefore have
$$|B_N(f, x_0) - f(x_0)| \leq \frac{M}{2\sqrt{N}} + \epsilon_N(x_0)$$

(a) If $f(x)$ is in $C[0, 1]$, it is uniformly continuous and $\epsilon_N(x_0)$ goes to zero uniformly as N approaches infinity. Thus part (a) is proven.

(b) If x_0 is a point of continuity of $f(x)$, then $\epsilon_N(x_0)$ and $|B_N(f, x_0) - f(x_0)|$ converge to zero as N approaches infinity, thereby proving part (b). ∎

Now assume that $|f(x) - f(y)| \leq |x - y|$ for all $x, y \in [0, 1]$. Then
$$\epsilon_N(x_0) \leq \frac{1}{N^{1/4}}.$$

Therefore
$$\left| B_N(f, x_0) - f(x_0) \right| \leq \frac{M}{2\sqrt{N}} + \frac{1}{N^{1/4}} \leq \frac{M_1}{N^{1/4}}$$

where M_1 is some constant.

Exercise. In this proof we used $N^{1/4}$ and ended up with a bound of $M/N^{1/4}$. Use N^α instead of $N^{1/4}$ and see if you can get a better bound.

THEOREM 5. *If $|f(x) - f(y)| \leq |x - y|$ for all $x, y \in [0, 1]$, then $|B_N(f, x_0) - f(x_0)| \leq 5/(4\sqrt{N})$ for all $x_0 \in [0, 1]$ and all N.*

Proof. Fix $x_0 \in [0, 1]$ and let x vary in $[0, 1]$. If
$$|x - x_0| \geq \frac{1}{\sqrt{N}}$$
then

$$|f(x) - f(x_0)| \leq |x - x_0| = \frac{|x - x_0|\sqrt{N}}{\sqrt{N}}$$

$$\leq \frac{|x - x_0|^2 N}{\sqrt{N}} < \frac{1 + |x - x_0|^2 N}{\sqrt{N}}.$$

It is obviously true that if

$$|x - x_0| \leq \frac{1}{\sqrt{N}}$$

then

$$|f(x) - f(x_0)| \leq \frac{1 + |x - x_0|^2 N}{\sqrt{N}}.$$

Therefore, the inequality holds for all x and $x_0 \in [0, 1]$. Thus,

$$|B_N(f, x_0) - f(x_0)|$$

$$= \left| \sum_{n=0}^{N} \left[f\left(\frac{n}{N}\right) - f(x_0) \right] \binom{N}{n} x_0^n (1 - x_0)^{N-n} \right|$$

$$\leq \sum_{n=0}^{N} \frac{1 + [(n/N) - x_0]^2 N}{\sqrt{N}} \binom{N}{n} x_0^n (1 - x_0)^{N-n}$$

$$= \frac{1}{\sqrt{N}} + \frac{x_0(1 - x_0)}{\sqrt{N}} \leq \frac{5}{4\sqrt{N}}. \blacksquare$$

Now that we have proven that we can come within $5/4\sqrt{N}$ for arbitrary $f(x)$ such that $|f(x) - f(y)| \leq |x - y|$, how do we know that we cannot find a bound which goes to zero even faster? To prove that we cannot, in general, do better, it suffices to prove that for one particular $f(x)$, in our class, we cannot do better.

We consider $f(x) = |x - 1/2|$. It is trivial that

$$|f(x) - f(y)| \leq |x - y|$$

for all x and $y \in [0, 1]$. To establish the lower bound it clearly will suffice to prove $\sqrt{N} |B_N(f, 1/2) - f(1/2)| \geq C > 0$.

LEMMA 6.

$$B_{2M}\left(f, \frac{1}{2}\right) = \frac{1}{2^{2M+1}} \binom{2M}{M}$$

while

$$B_{2M+1}\left(f, \frac{1}{2}\right) = \frac{1}{2^{2M+1}} \binom{2M}{M}$$

Proof. We will use the two trivial relationships

$$\binom{n}{k} = \binom{n}{n-k} \text{ and } \binom{n}{k}k = \binom{n}{n-k+1}(n-k+1)$$

$$B_{2M}\left(f, \frac{1}{2}\right) = \sum_{n=0}^{M}\left(\frac{1}{2} - \frac{n}{2M}\right)\binom{2M}{n}\frac{1}{2^{2M}}$$

$$+ \sum_{n=M+1}^{2M}\left(\frac{n}{2M} - \frac{1}{2}\right)\binom{2M}{n}\frac{1}{2^{2M}}$$

$$= \frac{1}{2^{M+1}}\left[\sum_{n=0}^{M}\binom{2M}{n} - \sum_{n=M+1}^{2M}\binom{2M}{n}\right]$$

$$+ \frac{1}{2^{2M}2M}\left[-\sum_{n=0}^{M}\binom{2M}{n}n + \sum_{n=M+1}^{2M}\binom{2M}{n}n\right]$$

$$= \frac{1}{2^{2M+1}}\left[\sum_{n=0}^{M}\binom{2M}{n} - \sum_{n=M+1}^{2M}\binom{2M}{2M-n}\right] + \frac{1}{2^{2M}2M}$$

$$\times \left[-\sum_{n=0}^{M}\binom{2M}{n}n + \sum_{n=M+1}^{2M}\binom{2M}{2M-n+1}2M-n+1\right]$$

$$= \frac{1}{2^{2M+1}}\left[\sum_{n=0}^{M}\binom{2M}{n} - \sum_{j=0}^{M-1}\binom{2M}{j}\right]$$

$$+ \frac{1}{2^{2M}2M}\left[-\sum_{n=0}^{M}\binom{2M}{n}n + \sum_{j=1}^{M}\binom{2M}{j}j\right]$$

$$= \frac{1}{2^{2M+1}}\binom{2M}{M}$$

$$B_{2M+1}\left(f, \frac{1}{2}\right) = \sum_{n=0}^{M}\left(\frac{1}{2} - \frac{n}{2M+1}\right)\binom{2M+1}{n}\frac{1}{2^{2M+1}}$$

$$+ \sum_{m=M+1}^{2M+1}\left(\frac{n}{2M+1} - \frac{1}{2}\right)\binom{2M+1}{n}\frac{1}{2^{2M+1}}$$

$$= \frac{1}{2^{2M+2}}\left[\sum_{n=0}^{M}\binom{2M+1}{n} - \sum_{n=M+1}^{2M+1}\binom{2M+1}{n}\right]$$

$$+ \frac{1}{2^{2M+1}(2M+1)}$$

$$\times \left[-\sum_{n=0}^{M}\binom{2M+1}{n}n + \binom{2M+1}{n}n\right]$$

$$= \frac{1}{2^{2M+2}}\left[\sum_{n=0}^{M}\binom{2M+1}{n} - \sum_{n=M+1}^{2M+1}\binom{2M+1}{2M+1-n}\right]$$

$$+ \frac{1}{2^{2M+1}(2M+1)}$$

$$\times \left[-\sum_{n=0}^{M} \binom{2M+1}{n} n \right.$$
$$\left. + \sum_{n=M+1}^{2M+1} \binom{2M+1}{2M+2-n}(2M+2-n) \right]$$
$$= \frac{1}{2^{2M+2}} \left[\sum_{n=0}^{M} \binom{2M+1}{n} - \sum_{j=0}^{M} \binom{2M+1}{j} \right]$$
$$+ \frac{1}{2^{2M+1}(2M+1)}$$
$$\times \left[-\sum_{n=0}^{M} \binom{2M+1}{n} n + \sum_{j=1}^{M+1} \binom{2M+1}{j} j \right]$$
$$= \frac{1}{2^{2M+1}(2M+1)} \binom{2M+1}{M+1}(M+1) = \frac{1}{2^{2M+1}} \binom{2M}{M}. \blacksquare$$

THEOREM 7. There is an $f(x)$ such that $|f(x) - f(y)| \leq |x - y|$ for all x and $y \in [0, 1]$ and a point $x_0 \in [0, 1]$ such that
$$\sqrt{N} |B_N(f, x_0) - f(x_0)| \geq C > 0$$
for some C and all N.

Proof. We let $f(x) = |x - 1/2|$ and let $x_0 = 1/2$. Then
$$\sqrt{N} |B_N(f, 1/2) - f(1/2)| = \begin{cases} \frac{\sqrt{2M}}{2^{2M+1}} \binom{2M}{M} & N = 2M \\ \frac{\sqrt{2M+1}}{2^{2M+1}} \binom{2M}{M} & N = 2M+1 \end{cases}$$

By Stirling's formula, $(\sqrt{2M}/2^{2M+1})\binom{2M}{M}$ approaches $1/\sqrt{2\pi}$ as M (and hence N) approaches infinity and $(\sqrt{2M+1}/2^{2M+1})\binom{2M}{M}$ also, obviously, approaches $1/\sqrt{2\pi}$. Thus we can find a $C > 0$ such that
$$\sqrt{N} |B_N(f, 1/2) - f(1/2)| \geq C$$
for all N. $\blacksquare$

We thus see that the previous theorem is basically the best possible; i.e., $1/\sqrt{N}$ is the correct order of magnitude of the convergence of $B_N(f, x)$ to $f(x)$ for all $f(x)$ such that $|f(x) - f(y)| \leq |x - y|$.

The Weierstrass theorem told us that any continuous function on $[0, 1]$ could be uniformly approximated by polynomials. In our discussion of Bernstein polynomials we were mainly concerned with "How well do the Bernstein polynomials approximate the functions of a certain class?"

This is an example of the two main questions in approximation theory. That is, given a sequence of functions

$$\{f_1(x), f_2(x), \ldots, f_n(x), \ldots\}$$

and a class C of functions (all having the same domain) we ask

(1) Can every of $f \in C$ be approximated by finite linear combinations of $\{f_1, f_2, \ldots\}$?

(2) Assuming the answer to question (1) is yes, how closely do (all or only certain) linear combinations of $\{f_1, f_2, \ldots, f_n\}$ approximate an $f \in C$, particularly as n approaches infinity?

By taking a different sequence $\{g_1, g_2, \ldots\}$ or a different class C_1 of functions with a different common domain of definition or a different measure of nearness in approximation or changing the permissible linear combinations of $\{f_1, f_2, \ldots\}$ we have an infinite number of questions on the linear theory of approximation.

CHAPTER II

PRELIMINARIES

In studying the convergence of Bernstein polynomials, we proved that $|B_N(f, x) - f(x)| \leq 5/(4\sqrt{N})$ for all $x \in [0, 1]$ and all f such that $|f(x) - f(y)| \leq |x - y|$.

Letting $\mathscr{S} = \{f \mid |f(x) - f(y)| \leq |x - y|\}$ we can restate the theorem as: $\|B_N(f, x) - f(x)\| \leq 5/(4\sqrt{N})$ for all $f \in \mathscr{S}$.

A weaker form of this theorem is: given any $f \in \mathscr{S}$ there is a polynomial $P(x)$ of degree N such that $\|P - f\| \leq 5/(4\sqrt{N})$.

We notice that the upper bound of $5/(4\sqrt{N})$ is independent of f. However, if we take arbitrary $f \in C[0, 1]$, then it is fairly obvious that the bound will have to be a function of f. (Just take an f which is not a polynomial so that $\|f - P\|$ is never zero and consider approximating $kf(x)$ as k approaches infinity.)

In a later chapter we will find a bound in terms of the following function of f.

(1) Moduli of Continuity

Definition. Let $f(x)$ be defined for $x \in [a, b]$. For each δ in $[0, b - a]$ let $\omega(f, \delta) = \sup_{|x-y|<\delta} |f(x) - f(y)|$. Then $\omega(f, \cdot)$ is called the modulus of continuity of $f(x)$.

Examples. (1) Let $f(x) \equiv 1$. Then $\omega(1, \delta) = \sup_{|x-y|<\delta} |1 - 1| = 0$. Therefore $\omega(1, \delta) \equiv 0$.

(2) Let $f(x) = x$. Then $\omega(x, \delta) = \sup_{|x-y|\leq\delta} |x - y| = \delta$. Therefore $\omega(x, \delta) = \delta$.

(3) Let $f(x) = x^2$ on $[0, 1]$. Then $\omega(x^2, \delta) = \sup |x^2 - y^2| = \sup |x - y| |x + y| \leq 2\delta$.

THEOREM 1. *If $\omega(f, \delta)/\delta$ goes to zero as δ approaches zero, then f is constant.*

Proof. Take $x_0 \in (a, b)$ and $0 < \delta < b - x_0$. Then

$$\frac{|f(x_0 + \delta) - f(x_0)|}{\delta} \leq \frac{\omega(f, \delta)}{\delta}$$

which approaches zero. Therefore, $f'(x_0) = 0$ for all x_0 and f is constant.

We note that $f(x)$ is continuous iff $\omega(f, 0^+) = 0$.

Let $f(x)$ be defined on $[0, b]$. Then $\omega(f, \delta)$ has the following properties (as a function of δ, $\delta \in [0, b]$).

(1) It is non-decreasing.
(2) It is subadditive, i.e., $\omega(f, \xi + \eta) \leq \omega(f, \xi) + \omega(f, \eta)$.
(3) Its value at 0 is 0.

Proof. Properties (1), and (3) are obvious.

(2) For each x and y such that $|x - y| \leq \xi + \eta$ we can find a z such that $|x - z| \leq \xi$ and $|z - y| \leq \eta$. Then

$$|f(x) - f(y)| \leq |f(x) - f(z)| + |f(z) - f(y)|$$
$$\leq \omega(f, \xi) + \omega(f, \eta).$$

Therefore

$$\sup_{|x-y| \leq \xi+\eta} |f(x) - f(y)| \leq \omega(f, \xi) + \omega(f, \eta). \blacksquare$$

In talking about the modulus of continuity of a periodic function with period 2π we can restrict δ to $[0, \pi]$. That is because we identity x and $2n\pi + x$ and thus no two points can be more than distance π apart.

If we are given a function $\omega_1(x)$ defined on $[a, c]$ which is non-decreasing and subadditive and such that $\omega_1(a) = 0$, we can define $\omega(x) = \omega_1(x + a)$ for $x \in [0, c - a]$ with properties (1), (2), and (3).

THEOREM 2. *Let $\omega(x)$ be defined on $[0, b]$ such that $\omega(x)$ has properties (1), (2), and (3). Then $\omega(x) = \omega(\omega, x)$.*

Proof. We must prove

$$\omega(\delta) = \sup_{|x-y| \leq \delta} |\omega(x) - \omega(y)|.$$

By taking $x = 0$ and $y = \delta$ we get

$$\omega(\delta) \leq \sup_{|x-y| \leq \delta} |\omega(x) - \omega(y)|.$$

Now assume we have an x and y such that $x + \delta \geq y > x$. Then

$$0 \leq \omega(y) - \omega(x) \leq \omega(y - x) \leq \omega(\delta).$$

Therefore, $|\omega(y) - \omega(x)| \leq \omega(\delta)$ and we are done by the symmetry in x and y. ∎

We can therefore state the following:

THEOREM 3. *A function $\omega(x)$ defined on $[0, b]$ is a modulus of continuity of a function on $[0, b]$ iff $\omega(x)$ has properties (1), (2), and (3).*

(2) Compact and Pre-Compact Sets

Definition. Let S be a set contained in a normed space. Then

(1) S is said to be pre-compact if every sequence of points from S contains a Cauchy convergent subsequence.

(2) S is said to be compact if every sequence of points from S contains a subsequence which converges to a point of S.

Here we will deal with the normed space $C[0, 1]$ (or $C^*[-\pi, \pi]$) which is a Banach space; *i.e.*, Cauchy convergence implies convergence to a member of $C[0, 1]$ (or $C^*[-\pi, \pi]$). Also f_n coverging to f in the topology of $C[0, 1]$ (or $C^*[-\pi, \pi]$) means f_n converges to f uniformly.

Definition. Let $\omega(\delta)$ be a continuous modulus of continuity. Then we define:

(a) $C_{\omega(\delta)} = \{f$ defined on $[0, 1]$ with $\omega(f, \delta) \leq \omega(\delta)\}$.
(b) $C_{\omega(\delta), M} = \{f \in C_{\omega(\delta)}$ with $|f(0)| \leq M\}$.

What we have called $\mathscr{S}$ is just C_δ.

Definition. Functions whose moduli of continuity are bounded by the same continuous modulus of continuity are called equicontinuous.

For completeness, we state here the famous Ascoli-Arzela theorem: A uniformly bounded, equicontinuous sequence of functions on $[a, b]$ contains a uniformly convergent subsequence.

THEOREM 4. $C_{\omega(\delta), M}$ *is compact.*

Proof. We need the boundedness since we can choose the sequence $\{f_1(x), f_2(x), \ldots\}$ with $f_n(x) \equiv n$ and we would not have a convergent subsequence, and yet $f_n(x) \in C_{\omega(\delta)}$.

Let $\{f_1(x), f_2(x) \ldots\}$ be an arbitrary sequence from $C_{\omega(\delta), M}$; *i.e.*, $\omega(f_n, \delta) \leq \omega(\delta)$ and $|f_n(0)| \leq M$. Then for any $x \in [0, 1]$,

$$|f_n(x)| \leq |f_n(x) - f_n(0)| + |f_n(0)| \leq \omega(f_n, x) + M$$
$$\leq \omega(f_n, 1) + M$$
$$\leq \omega(1) + M.$$

Therefore, the sequence is uniformly bounded. By Ascoli-Arzela $\{f_n(x)\}$

has a convergent subsequence $\{f_{n_k}(x)\}$ which converges to $f(x)$. Then we must show $f \in C_{\omega(\delta),\, M}$. It is obvious $|f(0)| \leq M$. Now choose any $\varepsilon > 0$ and δ, x, and y in $[0, 1]$ such that $|x - y| \leq \delta$. Then

$$|f(x) - f(y)| \leq |f(x) - f_{n_k}(x)| + |f_{n_k}(x) - f_{n_k}(y)|$$
$$+ |f_{n_k}(y) - f(y)|$$
$$\leq \varepsilon + \omega(f_{n_k}, \delta) + \varepsilon \text{ for large } n_k$$
$$\leq 2\varepsilon + \omega(\delta).$$

Since ε is arbitrary, we conclude that

$$|f(x) - f(y)| \leq \omega(\delta) \text{ or } \omega(f, \delta) \leq \omega(\delta). \blacksquare$$

Exercise. Prove: Let S be a pre-compact subset of $C[0, 1]$. Then $S \subseteq C_{\omega(\delta),\, M}$ for some $\omega(\delta)$ and some M.

Definition. $\mathscr{P}_n$ is the space of all real polynomials of degree at most n.

Definition. Let $f(x)$ be defined on $[0, 1]$. Then

$$E_n(f) = \inf_{p \in \mathscr{P}_n} \| f - p \| = \inf_{p \in \mathscr{P}_n} \sup_{x \in [0,1]} |f(x) - p(x)|$$

Definition. Let C be a set of functions on $[0, 1]$. Then

$$E_n(C) = \sup_{f \in C} E_n(f)$$

From the theory of Bernstein polynomials we know $E_N(\mathscr{S}) \leq 5/(4\sqrt{N})$ which goes to zero. In the next theorem we shall find a sufficient condition on a subset C of $C[0, 1]$ so that $E_n(C)$ converges to zero.

THEOREM 5. *Let C be a precompact subset of $C[0, 1]$. Then $E_n(C)$ converges to zero.*

Proof. If $E_n(C)$ does not go to zero, then, since $E_n(C) \geq E_{n+1}(C)$, there is an $\varepsilon > 0$ such that $E_n(C) > \varepsilon$ for all n. Therefore, for each n there is an $f_n(x) \in C$ with $E_n(f_n) > \varepsilon$.

Since C is precompact, there is a subsequence $\{f_{n_k}\}$ which converges uniformly to a function $f \in C[0, 1]$. By the Weierstrass theorem, for each n_k there is a $P_{n_k} \in \mathscr{P}_{n_k}$ with $\|f - P_{n_k}\|$ going to zero as n_k goes to infinity. Then

$$\|f_{n_k} - P_{n_k}\| \leq \|f_{n_k} - f\| + \|f - P_{n_k}\|$$

which goes to zero as n_k goes to infinity. But

$$0 < \varepsilon < E_{n_k}(f_{n_k}) \leq \|f_{n_k} - P_{n_k}\|$$

which is a contradiction. $\blacksquare$

Corollary. $E_n(\mathscr{S})$ converges to zero (even though $\mathscr{S}$ is not precompact).

Proof. For each $f \in \mathscr{S}$, $f(x) - f(0) \in C_{\delta,\,0}$ which is compact and therefore $E_n(C_{\delta,\,0})$ converges to 0. Since $\|(P_n(x) + f(0)) - f(x)\| = \|P_n(x) - (f(x) - f(0))\|$, we get $E_n(f(x) - f(0)) = E_n(f)$. Therefore $E_n(\mathscr{S}) = E_n(C_{\delta,\,0})$ which goes to zero.

Exercise. Prove $E_n(C_{\omega(\delta)})$ goes to zero for any $\omega(\delta)$.

Exercise. Find $E_0(\mathscr{S})$ and $E_1(\mathscr{S})$.

(3) Zeroes of Trigonometric Polynomials

We know that an n-th degree polynomial cannot have more than n zeroes (counting multiplicity). However, how many zeroes can an n-th degree trigonometric polynomial have? To make the question meaningful, we have to restrict the question to zeroes on $[0, 2\pi)$.

Theorem 6. Let $T_N(x)$ be a non-identically zero trigonometric polynomial of degree N. Then $T_N(x)$ has at most $2N$ zeros on $[0, 2\pi)$ (even counting multiplicities).

Proof. $T_N(x)$ can be written in the from $\sum_{n=-N}^{N} c_n e^{inx}$. Then $T_N(x)e^{iNx} = \sum_{n=0}^{2N} c_{n-N} e^{inx}$. Let $P(z) = \sum_{n=0}^{2N} c_{n-N} z^n$. Then since $P(z)$ has at most $2N$ zeroes (counting multiplicities) we are done as soon as we show that if x_0 is a zero of degree k of $T_N(x)$, then e^{ix_0} is a zero of degree k of $P(z)$.

Since $P(e^{ix_0}) = T_N(x_0)e^{iNx_0}$, it is obvious that e^{ix_0} is a root of $P(z)$ if x_0 is a root of $T_N(x)$. Also

$$P'(e^{ix})ie^{ix} = T_N(x)Nie^{iNx} + e^{iN_0}T'_N(x) \left\{ \text{where } P' = \frac{dP}{dz} = \frac{dP}{de^{ix}} \right\}$$

$$= e^{iNx}[T'_N(x) + iNT_N(x)].$$

Therefore x_0 a double root of $T_N(x)$ implies e^{ix_0} is a double root of $P(z)$. It is obvious that this process continues if x_0 is a k-th root of $T_N(x)$ and we would prove e^{ix_0} is a k-th root of $P(z)$. ∎

Corollary. Let $T_1(x)$ and $T_2(x)$ be trigonometric polynomials of degree N which agree at $2N+1$ distinct points of $[0, 2\pi)$. Then $T_1(x) \equiv T_2(x)$.

Corollary. Let $T_1(x)$ and $T_2(x)$ be trigonometric polynomials of degree N which agree at $2N+1$ distinct points of $[-\pi, \pi)$. Then $T_1(x) \equiv T_2(x)$.

POLYNOMIAL APPROXIMATION

(4) Derivatives of Polynomials

THEOREM 7. If $T(x)$ is a trigonometric polynomial of N-th degree, then $\|T'\| \leq N\|T\|$.

Note. The bound is sharp as seen by taking $T(x) = \cos Nx$.
Before we prove the theorem, we need three lemmas.

LEMMA 8. If $T(x)$ is an even trigonometric polynomial of degree $N - 1$, then

$$\sum_{k=1}^{N} T\left(\frac{2k-1}{2N}\pi\right) \left[\frac{\sin\left(\frac{2k-1}{2N}\pi\right)(-1)^{k+1}\cos Nx}{N\left(\cos x - \cos\frac{2k-1}{2N}\pi\right)} \right] \equiv T(x).$$

Proof. Let

$$f(x) = \sum_{k=1}^{N} T\left(\frac{2k-1}{2N}\pi\right) \left[\frac{\sin\left(\frac{2k-1}{2N}\pi\right)(-1)^{k+1}\cos Nx}{N\left(\cos x - \cos\frac{2k-1}{2N}\pi\right)} \right]$$

We observe that

$$\left[\frac{\sin\left(\frac{2k-1}{2N}\pi\right)(-1)^{k+1}\cos\left(\frac{2j-1}{2}\pi\right)}{N\left(\cos\frac{2j-1}{2N}\pi\right) - \cos\left(\frac{2k-1}{2N}\pi\right)} \right]$$

is 0 if $k \neq j$ and is 1 if $j = k$ where $j = 1, 2, \ldots, N$. Therefore

$$f\left(\frac{2j-1}{2N}\pi\right) = T\left(\frac{2j-1}{2N}\pi\right), \quad j = 1, 2, \ldots, N.$$

Since both $f(x)$ and $T(x)$ are even functions of x, we get equality also for $j = 0, -1, -2, \ldots, -N + 1$. We leave as an exercise the proof that

$$\frac{\cos Nx}{\cos x - \cos\frac{2k-1}{2N}\pi}$$

is a trigonometric polynomial of degree $N - 1$. Therefore, $f(x)$ is a trigonometric polynomial of degree $N - 1$ and, since $T(x)$ is also of degree $N - 1$ and agrees with $f(x)$ at $2N$ points in $(-\pi, \pi)$ we have that $f(x) \equiv T(x)$. ∎

LEMMA 9. $\sin x \geq 2x/\pi$ on $[0, \pi/2]$.

Proof. Let $f(x) = \sin x - 2x/\pi$. Then $f(0) = f(\pi/2) = 0$. Since $f'(0) = 1 - 2/\pi > 0$, $f(x)$ is increasing as x goes away from zero. Thus $f(x) > 0$ in $(0, \varepsilon)$ for some ε. If $f(x)$ were ever negative in $[0, \pi/2]$, $f'(x)$ would have to be zero at least twice in $(0, \pi/2)$. That would imply that $f''(x)$ is zero at least once in $(0, \pi/2)$. But $f''(x) = -\sin x$ is never zero in $(0, \pi/2)$. Therefore, $f(x) \geq 0$ in $[0, \pi/2]$. ∎

LEMMA 10. If $T(x)$ is an odd N-th degree trigonometric polynomial, then

$$\left\| \frac{T(x)}{\sin x} \right\| \leq N \| T \|.$$

Proof. We leave as an exercise, the proof that $T(x)/\sin x$ is an even trigonometric polynomial of degree $N - 1$. We can therefore use lemma 8 and without loss of generality we assume $\| T \| = 1$.

(a) For $\pi/2N \leq x \leq \pi/2$

$$\left| \frac{T(x)}{\sin x} \right| \leq \frac{1}{|\sin x|} \leq \frac{1}{\sin \frac{\pi}{2N}} \leq N \text{ by lemma 9 } \left(\text{set } x = \frac{\pi}{2N}\right)$$

(b) For $\pi/2 \leq x \leq \pi - \pi/2N$

$$\left| \frac{T(x)}{\sin x} \right| \leq \frac{1}{|\sin x|} \leq \frac{1}{\left|\sin \left(\pi - \frac{\pi}{2N}\right)\right|} = \frac{1}{\sin \frac{\pi}{2N}} \leq N$$

(c) For $0 \leq x < \pi/2N$ or $\pi - \pi/2N < x \leq \pi$

$$\left| \frac{T(x)}{\sin x} \right| = \left| \sum_{k=1}^{N} \frac{T\left(\frac{2k-1}{2N}\pi\right)(-1)^{k+1} \cos Nx}{N\left(\cos x - \cos \frac{2k-1}{2N}\pi\right)} \right|$$

$$\leq \sum_{k=1}^{N} \left| \frac{\cos Nx}{N\left(\cos x - \cos \frac{2k-1}{2N}\pi\right)} \right|$$

$$= \left| \frac{\sin Nx}{\sin x} \right| \quad \left\{\text{using lemma 8 for } \frac{\sin Nx}{\sin x}\right\}$$

$$\leq N \quad \blacksquare.$$

Proof of theorem 7. Assume $\| T \| = 1$. Pick a point ξ and we will show $|T'(\xi)| \leq N$.

$$\frac{T(\xi + x) - T(\xi - x)}{2}$$

is an N-th degree odd trigonometric polynomial in x of norm less than or equal to 1. Therefore, by lemma 10,

$$\left| \frac{T(\xi + x) - T(\xi - x)}{2 \sin x} \right| \leq N$$

for all x, or

$$\left| \frac{1}{2} \left[\frac{T(\xi + x) - T(\xi)}{x} + \frac{T(\xi) - T(\xi - x)}{x} \right] \frac{x}{\sin x} \right| \leq N$$

for all x. As x goes to 0, the left hand side approaches $|T'(\xi)|$, and we have proven $|T'(\xi)| \leq N$. ∎

THEOREM 11. *Let $P(x)$ be an N-th degree polynomial and let $\|P\| = \max_{-1 \leq x \leq 1} |P(x)|$. Then $\|P'\| \leq N^2 \|P\|$.*

Note: Once again, the bound is sharp for if we let

$$P(x) = \cos(N \arccos x),$$

then

$$\|P\| = 1 \text{ and } P'(x) = -(\sin(N \arccos x))N/\sqrt{1 - x^2}.$$

If we set $y = \arccos x$, then $P'(x) = -N \sin(Ny)/\sin y$ and therefore $\|P'\| \leq N^2$ and $|P'(1)| = N^2$.

Proof of theorem 11. Assume $\|P\| = 1$. Let $P_1(\theta) = P(\cos \theta)$. Then P_1 is an N-th degree trigonometric polynomial and $\|P_1\| \leq 1$. By theorem 7, $\|P_1'(\theta)\| \leq N$ (where $P_1'(\theta) = dP_1(\theta)/d\theta$). But

$$P_1'(\theta) = -\sin \theta P'(\cos \theta)$$

(where $P'(\cos \theta) = dP(\cos \theta)/d \cos \theta$) and therefore

$$|\sin \theta \, P'(\cos \theta)/N| \leq 1.$$

Also, $\sin \theta P'(\cos \theta)/N$ is an odd trigonometric polynomial of degree N and, by lemma 10, $|P'(\cos \theta)/N| \leq N$. Therefore $|P'(\cos \theta)| \leq N^2$ or $|P'(x)| \leq N^2$. ∎

We have just seen that we can have a $P \in \mathscr{P}_N$ for which $\|P'\| = N^2 \|P\|$ (namely, the Tchebychev polynomial $T_N(x)$). We recall that the N^2 was achieved at the endpoints of the interval $[-1, 1]$. It turns out, however, that if we keep to the interior of the interval, we can do better. This is illustrated by the following

THEOREM 12. For each $P \in \mathscr{P}_N$, $|P^{(k)}(0)| \leq N^k \|P\|$.

Proof. Without loss of generality, we assume $\|P\| = 1$. If we let $P(x) = \hat{P}(0) + \hat{P}(1)x + \cdots + \hat{P}(N)x^N$ then what we have to prove is that $k! |\hat{P}(k)| \leq N^k$.

The crucial part in this proof is the following.

LEMMA 13. Let k and N be non-negative integers with $k \leq N$ and let $P \in \mathscr{P}_N$ with $\|P\| = 1$. Then, if N and k are both even or both odd, $|\hat{P}(k)| \leq |\hat{T}_N(k)|$, and if one of N and k is odd and the other even, $|\hat{P}(k)| \leq |\hat{T}_{N-1}(k)|$.

Proof. We assume that k is even. (If k is odd the proof is similar and is left as an exercise.) We let $M = N$ if N is even and $M = N - 1$ if N is odd. We assume, moreover, that there is a $P \in \mathscr{P}_N$ such that $|\hat{P}(k)| > |\hat{T}_M(k)|$. Therefore there is a C such that $|C| < 1$ and $C\hat{P}(k) = \hat{T}_M(k)$. We then consider $T_M(x) - CP(x)$. This is in $\mathscr{P}_N$, has no x^k term, is positive when $T_M(x) = 1$ and is negative when $T_M(x) = -1$. We consider two cases: (a) If $M/2$ is even, then $T_M(\pm x_i) = 1$, and $T_M(\pm y_i) = -1$ where

$$0 = x_0 < y_1 < x_1 < y_2 < \cdots < y_{M/4} < x_{M/4} \leq 1.$$

Since $T_M(\pm x_i) - CP(\pm x_i) > 0$, we conclude that the even part of $T_M(x) - CP(x)$ is also positive at $\pm x_i$. Similarly, the even part of $T_M(x) - CP(x)$ is negative at $\pm y_i$. We can therefore conclude that the even part of $T_M(x) - CP(x)$ has at least $M/2$ zeroes in $(0, 1)$. However, since $T_M(x) - CP(x)$ has no x^k term, the even part of $T_M(x) - CP(x)$ involves at most $M/2$ terms and by Descarte's rule of signs such a polynomial cannot have $M/2$ positive zeroes. Thus there cannot be a $P \in \mathscr{P}_N$ such that $|\hat{P}(k)| > |\hat{T}_M(k)|$ while $\|P\| = 1$.

(b) If $M/2$ is odd, then $T_M(\pm x_i) = 1$ and $T_M(\pm y_i) = -1$ where $0 = y_1 < x_1 < y_2 < \cdots < y_{M/4+1/2} < x_{M/4+1/2} \leq 1$. As before, we conclude that the even part of $T_M(x) - CP(x)$ has at least $M/2$ positive zeroes which is impossible for a polynomial with at most $M/2$ terms. ∎

We now continue the proof of theorem 12, that $k! |\hat{P}(k)| \leq N^k$, when $P \in \mathscr{P}_N$. By lemma 13, $|P(x)|$ is either smaller than $|\hat{T}_N(k)|$ (if k and N are both even or both odd) or smaller than $|\hat{T}_{N-1}(k)|$ (otherwise). In Chapter I, we proved that

$$|\hat{T}_N(k)| = 2^k \frac{N+k}{N} \binom{\frac{N+k}{2}}{k}$$

if k and N are both even or both odd. Thus in the first case we have to prove that

$$k!\, 2^k \frac{N}{N+k} \binom{\frac{N+k}{2}}{k} \leq N^k$$

and in the second case we have to prove

$$k!\, 2^k \frac{N-1}{N-1+k} \binom{\frac{N-1+k}{2}}{k} \leq N^k.$$

However, the second case easily follows once the first inequality is proven and the first inequality is easily proven by expanding the binomial coefficient. ∎

(5) Limits of Polynomials

Suppose we are given a uniformly convergent sequence of polynomials all of degree less than or equal to some fixed number k. Is it true that the limit function will also be a polynomial of degree less than or equal to k? We can also ask this question about trigonometric polynomials. To answer these (and even more general questions) we need the following well known theorem, which we prove for completeness.

THEOREM 14. *In any finite dimensional vector space V any two norms are equivalent; i.e., if $\|\ \|_1$ and $\|\ \|_2$ are any two norms on V, then there are $A, B > 0$ with $A\|f\|_1 \leq \|f\|_2 \leq B\|f\|_1$ for all $f \in V$.*

Proof. It is trivial to prove that this is an equivalence relationship. Therefore, it will suffice to prove that any norm is equivalent to one fixed norm.

Let $V_1, \ldots, V_n$ be a basis for V. Then, every $f \in V$ is uniquely represented as $a_1 V_1 + \cdots + a_n V_n$. Let $\|f\| = \max |a_k|$. (This is the ℓ^∞ norm for V relative to $V_1, \ldots, V_n$).

We will now prove, that an arbitrary norm, $\|\cdot\|_1$, is equivalent to $\|\cdot\|$. First, we have that

$$\begin{aligned}\|f\|_1 &= \|a_1 V_1 + \cdots + a_n V_n\|_1 \\ &\leq |a_1|\|V_1\|_1 + \cdots + |a_n|\|V_n\|_1 \\ &\leq nM\|f\| \quad (\text{where } M = \max_k \|V_k\|_1).\end{aligned}$$

Now assume there is no A such that $A\|f\| \leq \|f\|_1$ for all $f \in V$.

Therefore, for each integer m, there is an $f_m \in V$ with $\|f_m\|_1 < (1/m)\|f_m\|$. Without loss of generality, we can assume that $\|f_m\| = 1$. We therefore have a sequence $\{f_m\}$ with $f_m \in V$, $\|f_m\| = 1$ and $\|f_m\|_1 < 1/m$. We can write $f_m = a_{1m}V_1 + \cdots + a_{nm}V_n$ with $|a_{im}| \leq 1$ and for each m, there is an i with $|a_{im}| = 1$. Since there are an infinite number of m and only a finite number of i's, at least one i must be repeated infinitely often and, without loss of generality, we assume that it is 1.

By extracting a subsequence $\{m_k\}$ we can have $f_{m_k} \in V$, $\|f_{m_k}\|_1 < 1/m_k$ and $f_{m_k} = a_{1m_k}V_1 + \cdots + a_{nm_k}V_n$ with $|a_{1m_k}| = 1$ and $|a_{im_k}| \leq 1$. By extracting another subsequence, we can assume that a_{1m_k} converges to a_1 with $|a_1| = 1$. By further extracting subsequences with can assume that a_{2m_k} converges to $a_2, \ldots$ and a_{nm_k} converges to a_n with $|a_i| \leq 1$.

Let $f = a_1V_1 + \cdots + a_nV_n$. Then $\|f_{m_k} - f\|_1 \leq nM\|f_{m_k} - f\|$ which goes to zero and therefore $\|f_{m_k} - f\|_1$ goes to zero. Then $\|f\|_1 \leq \|f - f_{m_k}\|_1 + \|f_{m_k}\|_1$ which goes to zero. Therefore $\|f\|_1 = 0$ and $f \equiv 0$. However that implies that $a_i = 0$ $i = 1, 2, \ldots, n$, which is impossible since $|a_1| = 1$.

THEOREM 15. Let $\varphi_1, \ldots \varphi_n$ be a set of linearly independent functions with a common domain. Let V be the vector space spanned by $\{\varphi_1, \ldots, \varphi_n\}$ and let $\|\ \|_1$ be a norm on V. Then, if $\{P_i\}$ is a Cauchy convergent sequence in V, there is a $P \in V$ with $\|P_i - P\|_1$ converging to zero.

Proof. We are given that $\|P_i - P_j\|_1$ goes to zero as i, j approach infinity. Let $P_i = a_{1i}\varphi_1 + \cdots a_{ni}\varphi_n$, and let $\|P_i\| = \max_k |a_{ki}|$. Then by the previous theorem, we have $\|P_i - P_j\|$ approaching zero as i, j approach infinity. Then, for each k, $\{a_{kj}\}$ is Cauchy convergent and hence convergent. Assume a_{kj} converges to a_k and let

$$P = a_1\varphi_1 + \cdots + a_n\varphi_n.$$

Then, since $\|P_j - P\|$ goes to 0, we have $\|P_j - P\|_1$ approaching zero.

Corollary. The uniform limit of polynomials of degree less than or equal to k, is a polynomial of degree less than or equal to k.

Corollary. The uniform limit of trigonometric polynomials of degree less than or equal to k, is a trigonometric polynomial of degree less than or equal to k.

CHAPTER III

POLYNOMIALS OF BEST APPROXIMATION

In this chapter, we concern ourselves with questions about "best approximation" and in particular with the questions of existence, characterization and uniqueness of polynomials of best approximation. We first deal with algebraic polynomials and then, by mimicking the previous proofs, with trigonometric polynomials. The theorems will be stated and proven for functions on [0, 1] but it is obvious that they apply equally well to $[a, b]$ where $-\infty < a < b < \infty$.

(1) Existence

THEOREM 1. Let $f \in C[0, 1]$. Then there is a $P \in \mathscr{P}_n$ such that

$$\|f - P\| = \inf_{P_n \in \mathscr{P}_n} \|f - P_n\| = E_n(f)$$

i.e. a best n-th degree approximating polynomial exists.

Proof. Let $T = \{P \in \mathscr{P}_n \mid \|P\| \leq 2\|f\|\}$. If $P \in \mathscr{P}_n$ but $\|P\| > 2\|f\|$, then $\|P - f\| \geq \|P\| - \|f\| > \|f\|$. Thus P would be a worse approximant than 0. We therefore have that

$$\inf_{P \in \mathscr{P}_n} \|f - P\| = \inf_{P \in T} \|f - P\| = E_n(f).$$

Since for $P \in T$, $|P(0)| \leq 2\|f\|$ and

$$\omega(P, \delta) \leq \|P'\|\delta \leq n^2\|P\|\delta \leq 2n^2\|f\|\delta,$$

we have $T \subseteq C_{2n^2\|f\|\delta, 2\|f\|}$.

Since $C_{2n^2\|f\|\delta, 2\|f\|}$ is compact, from any sequence from T, we can extract a uniformly convergent subsequence. However, that subsequence will be a uniformly convergent sequence from $\mathscr{P}_n$ and hence its limit will also be in $\mathscr{P}_n$. Since, it is obvious that the absolute value of the limit function evaluated at 0 cannot be more than $2\|f\|$, the limit function will be in T and T is compact. In particular, let $\{P_k\}$ be a sequence from T such that $\|f - P_k\|$ converges to $E_n(f)$. We can extract a convergent subsequence $\{P_{n_k}\}$ which converges uniformly to a $P_0 \in T$. For a given $\varepsilon > 0$, there is an N such that $n_k > N$ implies $\|P_{n_k} - P_0\| < \varepsilon$ and $\|f - P_{n_k}\| \leq E_n(f) + \varepsilon$. Therefore, for $n_k > N$, $\|f - P_0\| \leq \|f - P_{n_k}\| + \|P_{n_k} - P_0\| \leq 2\varepsilon + E_n(f)$. Since ε is arbitrary, $\|f - P_0\| \leq E_n(f)$ and is obviously equal to it. ∎

Note. Actually, all we need in this theorem is for f to be bounded.

(2) Characterization

Assume $f \in C[0, 1]$ with $M = \max_{x \in [0,1]} f(x)$ and $m = \min_{x \in [0,1]} f(x)$. Assume, moreover, that $f(x_1) = M$ and $f(x_2) = m$. If C is any real number smaller than $(M + m)/2$, then

$$f(x_1) - C = M - C > M - \frac{M+m}{2} = \frac{M-m}{2} \geq 0,$$

while if C is any real number greater than $(M + m)/2$, then

$$f(x_2) - C = m - C < m - \frac{M+m}{2} = \frac{m-M}{2} \leq 0.$$

Thus, C not equal to $(M + m)/2$ implies that $\|f - C\| > (M - m)/2$. On the other hand, if $(M + m)/2 \leq f(x) \leq M$, then

$$0 \leq f(x) - \frac{M+m}{2} \leq M - \frac{M+m}{2} = \frac{M-m}{2}$$

and if $m \leq f(x) \leq (M + m)/2$, then

$$0 \geq f(x) - \frac{M+m}{2} \geq m - \frac{M+m}{2} = -\frac{M-m}{2}.$$

Thus $\|f - (M+m)/2\| \leq (M-m)/2$ and $(M+m)/2$ is a better approximant to $f(x)$ than any other constant. Since

$$f(x_1) - \frac{M+m}{2} = \frac{M-m}{2}$$

and

$$f(x_2) - \frac{M+m}{2} = -\frac{M-m}{2}$$

we see that $\|f - (M+m)/2\| = (M-m)/2$ and this maximum difference is achieved at least twice, once positive and once negative.

Example. Prove $E_0(\mathscr{S}) = 1/2$. For each $f \in \mathscr{S}$,

$$E_0(f) = \frac{1}{2}[\max f - \min f].$$

There is an $x_1 \in [0, 1]$ and an $x_2 \in [0, 1]$ such that $f(x_1) = \max f(x)$ and $f(x_2) = \min f(x)$. Therefore,

$$E_0(f) = \frac{1}{2}[f(x_1) - f(x_2)] \le \frac{1}{2}|x_1 - x_2| \le \frac{1}{2}.$$

Thus

$$E_0(\mathscr{S}) = \sup_{f \in \mathscr{S}} E_0(f) \le \frac{1}{2}.$$

To prove $E_0(\mathscr{S}) = 1/2$, it will suffice to find one $f \in \mathscr{S}$ such that $E_0(f) = 1/2$. Let $f(x) = x$. Then $f(x) \in \mathscr{S}$ and $\max f = 1$ and $\min f = 0$. Thus,

$$E_0(f) = \frac{1}{2}[\max f - \min f] = \frac{1}{2}.$$

The property, that the maximum difference between a function $f(x)$ and a best constant approximant is achieved twice and with different signs, is one that we would like to generalize for n-th degree approximants. For this we need the following

Definition. A function $g \in C[0, 1]$ has the *n*-alternation property ($n \ge 1$) if we can find $\{x_i\}_{i=1}^n$ with $0 \le x_1 < x_2 < \cdots < x_n \le 1$ such that either

(a) $\qquad g(x_i) = (-1)^i \|g\|, \quad i = 1, 2, \cdots, n$

or

(b) $\qquad g(x_i) = (-1)^{i+1}\|g\|, \quad i = 1, 2, \cdots, n.$

Since for every $g \in C[0, 1]$, $\|g\|$ is achieved at least once, every $g \in C[0, 1]$ has the 1-alternation property. Also, by what we just saw, if $f \in C[0, 1]$, and C is a best constant approximant, then $f(x) - C$ has the 2-alternation property.

If a function $g \in C[0, 1]$ has the n-alternation property then it has the m-alternation property for $m = 1, 2, \cdots, n$. Therefore, if g does not have the n alternation property it does not have the m-alternation property if $m \ge n$.

THEOREM 2. Let $g \in C[0, 1]$. Then g has the n-alternation property but not the $n + 1$-alternation property iff there are n disjoint closed intervals $I_i \subseteq [0,1]\, i = 1, \cdots, n$ such that

(1) if $I_i = [a_i, b_i]$, then $a_1 < b_1 < a_2 < b_2 < \cdots < b_n$
(2) $\sup_{x \notin \bigcup_{i=1}^{n} I_i} |g(x)| < \|g\|$ and either

(a) $$\max_{x \in I_i} (-1)^i g(x) = \|g\|$$

and

$$\min_{x \in I_i} (-1)^i g(x) > -\|g\|, \quad i = 1, \ldots, n$$

or

(b) $$\max_{x \in I_i} (-1)^{i+1} g(x) = \|g\|$$

and

$$\min_{x \in I_i} (-1)^{i+1} g(x) > -\|g\|, \quad i = 1, \ldots, n.$$

The proof of this theorem, which easily follows from the properties of a continuous function, is left as an exercise.

We are now ready to state the generalization of our statement about best constant approximations. To interpret the theorem correctly for constant approximation we have to consider the degree of a constant function as being zero.

THEOREM 3. Let $f(x)$ be in $C[0, 1]$ and let $P_n(x)$ be a best n-th degree polynomial approximant to $f(x)$. Then $f(x) - P_n(x)$ has the $n+2$-alternation property.

Proof. $f(x) - P_n(x)$ must have the 1-alternation property since it is continuous. Assume the theorem is false and let m be the largest integer such that $f(x) - P_n(x)$ has the m-alternation property. By the previous theorem, we know there exists m disjoint closed intervals $I_i, i = 1 \cdots m$ such that

$$\sup_{x \notin \bigcup_{i=1}^{m} I_i} |f(x) - P_n(x)| < \|f - P_n\| = E_n(f)$$

and either property (a) or property (b) of that theorem holds.. Without loss of generality we can assume property (b) holds, *i.e.*,

$$\max_{x \in I_i} (-1)^{i+1} (f(x) - P_n(x)) = E_n(f)$$

and

28   POLYNOMIAL APPROXIMATION

$$\min_{x \in I_i} (-1)^{i+1} (f(x) - P_n(x)) > - E_n(f), \quad i = 1, 2, \cdots, m.$$

Let

$$C_1 = \sup_{x \notin \bigcup_{i=1}^{m} I_i} |f(x) - P_n(x)|$$

and let

$$C_2 = \min_i \min_{x \in I_i} (-1)^{i+1} (f(x) - P_n(x))$$

Then $C_1 < E_n(f)$ and $C_2 > - E_n(f)$. Now choose $C > 0$ such that $C < E_n(f) - C_1$ and $C < E_n(f) + C_2$. If we write $I_i = [\alpha_i, \beta_i]$, with $\beta_i < \alpha_{i+1}$, let $\xi_k = (1/2)[\alpha_{k+1} + \beta_k]$, $k = 1, 2 \ldots, m - 1$. Let $P(x) = C\pi_{k=1}^{m-1} (\xi_k - x)$. Then $\|P\| \leq C$, $(-1)^{i+1} P(x) > 0$ in I_i and $P(x)$ is a polynomial of degree $m - 1$ (since we are assuming $m \leq n + 1$, degree $P \leq n$).

Our theorem will be proven, if we can show that

$$\|f - P_n - P\| < E_n(f),$$

since $P_n(x) + P(x)$ is a polynomial of degree at most n.

Take $x \notin \bigcup_{i=1}^{m} I_i$. Then

$$\sup_{x \notin \bigcup_{i=1}^{m} I_i} |f(x) - P_n(x) - P(x)|$$

$$\leq \sup_{x \notin \bigcup_{i=1}^{m} I_i} |f(x) - P_n(x)| + \|P\|$$

$$\leq C_1 + C$$

$$< E_n(f).$$

Now take $x \in I_i$. Then

$$(-1)^{i+1}(f(x) - P_n(x) - P(x))$$
$$= (-1)^{i+1}(f(x) - P_n(x)) - (-1)^{i+1}P(x)$$
$$\geq C_2 - C > - E_n(f).$$

Also, since $-(-1)^{i+1} P(x) < 0$ in I_i, we have

$$(-1)^{i+1} (f(x) - P_n(x) - P(x))$$
$$= (-1)^{i+1}(f(x) - P_n(x)) - (-1)^{i+1}P(x)$$
$$< (-1)^{i+1}(f(x) - P_n(x)) \leq E_n(f).$$

Thus for any x in any I_i, we have

$$|f(x) - P_n(x) - P(x)| < E_n(f)$$

and since $\bigcup_{i=1}^{m} I_i$ is compact, we also have

$$\sup_{x \in \bigcup_{i=1}^{m} I_i} |f(x) - P_n(x) - P(x)| < E_n(f).$$

Combining this with

$$\sup_{x \notin \bigcup_{i=1}^{m} I_i} |f(x) - P_n(x) - P(x)| < E_n(f)$$

we get $\|f - P_n - P\| < E_n(f)$. ∎

Example. Find $E_1(\sqrt{x})$. Let $ax + b$ be a best linear approximation to $\sqrt{x}$. Then we have 3 points, x_1, x_2, x_3, such that

$$0 \le x_1 < x_2 < x_3 \le 1$$

and

$$\sqrt{x_1} - ax_1 - b = \pm E_1(\sqrt{x})$$
$$\sqrt{x_2} - ax_2 - b = \mp E_1(\sqrt{x})$$
$$\sqrt{x_3} - ax_3 - b = \pm E_1(\sqrt{x}).$$

Moreover, since $\sqrt{x} - ax - b$ has a maximum or minimum at the interior point x_2, we have $1/(2\sqrt{x_2}) - a = 0$.

If $x_1 \ne 0$, then x_1 is an interior point and we would have $1/2\sqrt{x_1} - a = 0$, which is impossible. Therefore $x_1 = 0$ and by a similar argument $x_3 = 1$. We thus have the following four equations

(1) $\qquad\qquad\qquad -b = \pm E_1(\sqrt{x})$

(2) $\qquad\qquad \sqrt{x_2} - ax_2 - b = \mp E_1(\sqrt{x})$

(3) $\qquad\qquad\qquad 1 - a - b = \pm E_1(\sqrt{x})$

(4) $\qquad\qquad\qquad \dfrac{1}{2\sqrt{x_2}} - a = 0$

Subtracting equation (1) from equation (3), we have $a = 1$. Thus $x_2 = 1/4$. Substituting for a and x_2 in equation (2) and adding equation (1) we get

$$\frac{1}{2} - \frac{1}{4} - 2b = 0, \text{ and } b = \frac{1}{8}.$$

Thus, $E_1(\sqrt{x}) = 1/8$, $x - 1/8$ is a best approximation and $0, 1/4$, and 1 are the 3 points from the alternation theorem.

Exercise. (a) Find $E_1(e^x)$; (b) find $E_2(\sqrt{x})$.

(3) Uniqueness

THEOREM 4. The best n-th degree approximant to a given function $f(x) \in C[0,1]$ is unique.

Proof. Assume $P_1(x)$ and $P_2(x)$ are 2 best n-th degree approximants to $f(x)$; i.e., $\|f - P_i\| = E_n(f)$, $i = 1, 2$. Then for any $x \in [0, 1]$

$$\left| f(x) - \frac{1}{2}[P_1(x) + P_2(x)] \right|$$

$$= \left| \frac{1}{2}[f(x) - P_1(x)] + \frac{1}{2}[f(x) - P_2(x)] \right|$$

$$\leq \frac{1}{2}\left| f(x) - P_1(x) \right| + \frac{1}{2}\left| f(x) - P_2(x) \right|$$

$$\leq E_n(f).$$

Therefore, $1/2\,(P_1(x) + P_2(x))$ is a best n-th degree approximant to $f(x)$. By theorem 3, we can find $n + 2$ points, $x_1, x_2, \ldots, x_{n+2}$ such that

$$\left| f(x_i) - \frac{1}{2}(P_1(x_i) + P_2(x_i)) \right| = E_n(f), \quad i = 1, 2, \ldots, n + 2,$$

We then have

$$E_n(f) = \left| \frac{1}{2}[f(x_i) - P_1(x_i)] + \frac{1}{2}[f(x_i) - P_2(x_i)] \right|$$

$$\leq \frac{1}{2}\left| f(x_i) - P_1(x_i) \right| + \frac{1}{2}\left| f(x_i) - P_2(x_i) \right|$$

$$\leq \frac{1}{2} E_n(f) + \frac{1}{2} E_n(f) = E_n(f).$$

Thus both sets of inequalities must be equalities. From the second set, we get

$$|f(x_i) - P_1(x_i)| = |f(x_i) - P_2(x_i)| = E_n(f), \quad i = 1, 2, \ldots, n + 2,$$

and from the first set of inequalities we get that

$$\text{sign}\,[f(x_i) - P_1(x_i)] = \text{sign}\,[f(x_i) - P_2(x_i)], \quad i = 1, 2, \ldots, n + 2.$$

Therefore, $P_1(x_i) = P_2(x_i)$, $i = 1, 2, \ldots, n + 2$. That, of course implies $P_1(x) \equiv P_2(x)$.

In theorem 3 we saw that if $f \in C[0, 1]$ and $P(x)$ is the best n-th degree approximant to $f(x)$ then $f(x) - P(x)$ has the $(n + 2)$-alternation property. In the next theorem we see that this property characterizes best approximations.

THEOREM 5. Let $f \in C[0, 1]$ and let $P(x)$ be a polynomial in $\mathscr{S}_n$ such that $f - P$ has the $(n + 2)$-alternation property. Then $P(x)$ is the best n-th degree approximant to $f(x)$ and $E_n(f) = \|f - P\|$.

Proof. Without loss of generality, we assume we have $x_1, x_2, \ldots, x_{n+2}$ such that $0 \leq x_1 < x_2 < \ldots < x_{n+2} \leq 1$ and

$$(-1)^i(f(x_i) - P(x_i)) = \|f - P\|, \quad i = 1, 2, \ldots, n + 2.$$

Let $P_n(x)$ be the best n-th degree approximant to $f(x)$. Assume $P_n(x)$ is a better n-th degree approximant to $f(x)$ than $P(x)$; i.e.,

$$\|f - P_n\| < \|f - P\|.$$

Then

$$(-1)^i(f(x_i) - P_n(x_i)) < (-1)^i(f(x_i) - P(x_i)), \quad i = 1, 2, \ldots, n + 2.$$

Subtracting $(-1)^i(f(x_i) - P_n(x_i))$ from both sides, we get

$$(-)^i(P_n(x_i) - P(x_i)) > 0, \quad i = 1, 2, \ldots, n + 2.$$

Thus $P_n(x) - P(x)$ changes signs at adjacent x_i and therefore has a zero between adjacent x_i. Altogether, we get at least $n + 1$ zeros for a polynomial in $\mathscr{S}_n$. Therefore, $P_n(x) \equiv P(x)$, and $P(x)$ is the best n-th degree approximant to $f(x)$.

Corollary. Let $f(x) \in C[0, 1]$ have the $(n + 2)$-alternation property. Then $E_0(f) = E_1(f) = \ldots = E_n(f) = \|f\|$.

Proof. We just apply the previous theorem with $P(x) \equiv 0$. ∎

Example. Find $E_n(x^{n+1})$ (for $x \in [-1, 1]$). What we are looking for is

$$\min_{a_0, a_1, \ldots, a_n} \|x^{n+1} - a_n x^n - \ldots - a_0\| \quad \text{where } \|f\| \text{ is } \sup_{x \in [-1,1]} |f(x)|.$$

Let $T_{n+1}(x) = (1/2^n) \cos((n + 1) \arccos x)$. (This is the Tchebychev polynomial multiplied by $1/2^n$)). It is easily seen that $T_{n+1}(x)$ is a polynomial of degree $n + 1$, the coefficient of x^{n+1} is 1 and $\|T_{n+1}(x)\| = 1/2^n$. Setting $x_k = \cos k\pi/(n + 1)$, $k = 0, 1, 2, \ldots, n + 1$, we see that

$$T_{n+1}(x_k) = \frac{(-1)^k}{2^n}, \quad k = 0, \ldots, n + 1$$

or

$$(-1)^k T_{n+1}(x_k) = \|T_{n+1}\|, \quad k = 0, \ldots, n + 1.$$

Thus T_{n+1} has the $(n + 2)$-alternation property and by the corollary, $E_n(T_{n+1}) = 1/2^n$. However, since $T_{n+1} = x^{n+1} + P(x)$, where $P(x) \in \mathscr{P}_n$, $E_n(T_{n+1}) = E_n(x^{n+1})$. Thus $E_n(x^{n+1}) = 1/2^n$.

We may also state this result as: of all monic polynomials of degree $n + 1$, the one with minimum norm is $T_{n+1}(x)$.

Example. Find $E_1(\mathscr{S})$. To find this, we have to find two things: first, a lower bound for one $f \in \mathscr{S}$ and, second, an upper bound for every $f \in \mathscr{S}$.

(1) Let $f(x)$ be defined as $f(0) = f(1) = 1/4 = -f(1/2)$ and linear in between. It is easily seen that $f \in \mathscr{S}$. Moreover, $f(x)$ has the 3-alternation property. Thus $E_1(f) = \|f\| = 1/4$. Since $E_1(\mathscr{S}) = \sup_{f \in \mathscr{S}} E_1(f)$ we get $E_1(\mathscr{S}) \geq 1/4$.

(2) Take any $f \in \mathscr{S}$ and assume $P_1(x)$ is the best linear approximant to $f(x)$. Then, by theorem 3, there are 3 points

$$x_1, x_2, x_3, \; 0 \leq x_1 < x_2 < x_3 \leq 1$$

such that

$$f(x_1) - P_1(x_1) = \pm E_1(f)$$
$$f(x_2) - P_1(x_2) = \mp E_1(f)$$
$$f(x_3) - P_1(x_3) = \pm E_1(f)$$

If we write $P_1(x) = mx + b$, we can rewrite these equations as

(1) $\qquad f(x_1) - mx_1 - b = \pm E_1(f)$
(2) $\qquad f(x_2) - mx_2 - b = \mp E_1(f)$
(3) $\qquad f(x_3) - mx_3 - b = \pm E_1(f)$

Subtracting equation (2) from (1) and equation (3) from (2), we get

$$f(x_1) - f(x_2) - m(x_1 - x_2) = \pm 2E_1(f)$$
$$f(x_2) - f(x_3) - m(x_2 - x_3) = \mp 2E_1(f)$$

Solving these two equations for m, we get

$$\pm E_1(f) = \dfrac{\dfrac{f(x_1) - f(x_2)}{x_1 - x_2} - \dfrac{f(x_2) - f(x_3)}{x_2 - x_3}}{\dfrac{2}{x_1 - x_2} + \dfrac{2}{x_2 - x_3}}$$

Since $f \in \mathscr{S}$, the absolute value of the numerator is ≤ 2. Thus

$$E_1(f) = |\pm E_1(f)| \leq \dfrac{1}{\dfrac{1}{x_2 - x_1} + \dfrac{1}{x_3 - x_2}}$$

$$= \frac{(x_2 - x_1)(x_3 - x_2)}{(x_3 - x_1)}$$

$$\leq \frac{(x_2 - x_1)(x_3 - x_2)}{(x_3 - x_1)^2}$$

Let $A = x_2 - x_1$ and $B = x_3 - x_2$. Then $E_1(f) \leq AB/(A + B)^2 \leq 1/4$ by using the inequality $(A - B)^2 \geq 0$. Thus $E_1(f) \leq 1/4$ for all $f \in \mathscr{S}$ and together with our lower bound from before we get $E_1(\mathscr{S}) = 1/4$.

Example. Prove $E_n(\mathscr{S}) \geq 1/2(n + 1)$.

Define $f(x)$ by having $f(k/n + 1) = (-1)^k/2(n + 1)$ and $f(x)$ linear between. It is easily seen that $f \in \mathscr{S}$ and $f(x)$ has the $(n + 2)$-alternation property. Thus

$$E_n(f) = \|f\| = \frac{1}{2(n + 1)}$$

and

$$E_n(\mathscr{S}) \geq E_n(f) = \frac{1}{2(n + 1)}.$$

Later, we shall see via Jackson's theorem that $E_n(\mathscr{S}) \leq A/(n + 1)$ where A is independent of n. Thus we shall know that $1/(n + 1)$ is the correct order of magnitude.

(4) Best Trigonometric Approximation

So far, in this chapter, we have dealt with approximation to continous functions on $[0, 1]$ (or $[a, b]$). Actually these theorems can also be stated for continuous 2π periodic functions with the same proofs. We therefore just state the corresponding theorems and leave the proofs to the reader.

Definition. $\mathscr{T}_n$ is the space of real trigonometric polynomials of degree at most n. We note that $\mathscr{T}_n$ is a vector space of dimension $2n + 1$.

Definition. For $f \in C^*[-\pi, \pi]$, $E_n^*(f) = \inf_{T \in \mathscr{T}_n} \|f - T\|$ where $\|g\|$ refers to $\sup_{x \in [-\pi, \pi]} |g(x)|$.

THEOREM 6. Let $f \in C^*[-\pi, \pi]$. Then there is a $T \in \mathscr{T}_n$ such that

$$\|f - T\| = \inf_{T_n \in \mathscr{T}_n} \|f - T_n\| = E_n^*(f)$$

Definition. A function $g \in C^*[-\pi, \pi]$ has the n-alternation property if we can find $\{x_i\}_{i=1}^n$ with $-\pi \leq x_1 < x_2 < \ldots < x_n < \pi$ such that either

(a) $\qquad g(x_i) = (-1)^i \|g\|, \quad i = 1, \ldots, n$

or

(b) $$g(x_i) = (-1)^{i+1}\|g\|, \quad i = 1, \ldots, n.$$

THEOREM 7. Let $f(x)$ be in $C^*[-\pi, \pi]$ and let $T_n(x)$ be a best n-th degree trigonometric polynomial approximant to $f(x)$. Then $f(x) - T_n(x)$ has the $(2n + 2)$-alternation property.

THEOREM 8. The best n-th degree trigonometric approximant to a given $f \in C^*[-\pi, \pi]$ is unique.

THEOREM 9. Let $f \in C^*[-\pi, \pi]$ and let $T(x)$ be $\in \mathscr{T}_n$ and such that $f - T$ has the $(2n + 2)$-alternation property. Then $T(x)$ is the best n-th degree trigonometric approximant to $f(x)$ and $E_n^*(f) = \|f - T\|$.

Corollary. Let $f \in C^*[-\pi, \pi]$ have the $(2n + 2)$-alternation property. Then $E_0^*(f) = E_1^*(f) = \ldots = E_n^*(f) = \|f\|$.

Example. Find $E_n^*(\sin(n+1)x)$. Let

$$x_1 = -\frac{-(2n+1)\pi}{2(n+1)}$$

and

$$x_{k+1} = -\frac{(2n+1)\pi}{2(n+1)} + \frac{k\pi}{n+1}$$

where $k = 1, \ldots, 2n + 1$. Then, if n is even, $\sin(n+1)x_k = (-1)^k$, $k = 1, 2, \ldots, 2n + 2$; whereas, if n is odd, $\sin(n+1)x_k = (-1)^{k+1}$, $k = 1, 2, \ldots, 2n + 2$. Thus, in either case, $\sin(n+1)x$ has the $(2n + 2)$-alternation property. By the Corollary,

$$E_n(\sin(n+1)x) = \|\sin(n+1)x\| = 1.$$

Definition. $\mathscr{S}^* = \{f \in C^*[-\pi, \pi] \text{ such that } \omega(f, \delta) \leq \delta\}$.

Definition. Let C be a class of periodic functions on $[-\pi, \pi]$. Then $E_n^*(C) = \sup_{f \in C} E_n^*(f)$.

EXAMPLE. $E_n^*(\mathscr{S}^*) \geq \pi/2(n+1)$.

Proof. Define $f(x)$ by $f(\pm k\pi/(n+1)) = (-1)^k \pi/2(n+1)$ for $k = 0, \ldots, (n+1)$ and $f(x)$ is linear in between.

Then it is easily seen that $f \in \mathscr{S}^*$, that $f(x)$ has the $(2n + 2)$-alternation property, and $E_n^*(f) = \|f\|$. Then

$$E_n^*(\mathscr{S}^*) \geq E_n^*(f) = \|f\| = \pi/(n+1).$$

CHAPTER IV

JACKSON'S THEOREMS

In the previous chapters, we have proven some theorems concerning lower bounds for the order of approximation. We showed an example of an $f \in \mathscr{S}$ such that $E_n(f) \geq 1/2n$ and thereby established that $E_n(\mathscr{S}) \geq 1/2n$. We proved a similar statement about $E_n^*(\mathscr{S}^*)$.

In this chapter, we will prove (among other things) that that is the correct order of magnitude for $E_n(\mathscr{S})$, inasmuch as we show that $E_n(\mathscr{S}) \leq C/n$ for some constant C (similarly for $E_n^*(\mathscr{S}^*)$),

We shall also derive an upper bound for $E_n(f)$ for any f and upper bounds for the classes of functions with higher degrees of smoothness (*i.e.*, more differentiability). All of these fall under the category of theorems known as Jackson's theorems (or, more correctly, Jackson-type theorems).

First, though, we will need to know some things about Gaussian quadrature, all of which we derive here.

(1) Gaussian Quadrature

We would like to find $\{C_i\}_{i=1}^n$ and $\{\xi_i\}_{i=1}^n$ ($|\xi_i| \leq 1$) such that $\int_{-1}^1 P(x)\,dx = C_1 P(\xi_1) + \cdots + C_n P(\xi_n)$ for all $P \in \mathscr{P}_{2n-1}$. This is called the Gaussian quadrature formula. Assume there exist such $\{C_i\}$ and $\{\xi_i\}$. What would they have to be?

Let $\pi(x) = \prod_{i=1}^n (x - \xi_i)$ and let k be a non-negative integer $\leq n-1$. Then $\int_{-1}^1 x^k \pi(x)\,dx = \sum_{i=1}^n C_i \xi_i^k \pi(\xi_i) = 0$. Therefore, $\pi(x)$ is orthogonal to all of $\mathscr{P}_{n-1}$. Therefore, $\pi(x)$ is the Legendre polynomial and the $\{\xi_i\}_{i=1}^n$ are the zeros of the Legendre polynomial of degree n.

Now consider $\int_{-1}^1 \pi(x)/(x-\xi_k)\,dx$. It is obviously equal to $C_k \pi'(\xi_k)$. Thus the C_k are determined. All we will need to know about C_k, that $C_k > 0$, can be gotten by considering $Q_k(x) = \prod_{i=1, i \neq k}^n (x - \xi_1)^2$. Then $\int_{-1}^1 Q_k(x)\,dx = C_k Q_k(\xi_k)$, and since $\int_{-1}^1 Q_k(x)\,dx > 0$ and $Q_k(\xi_k) > 0$, we get that $C_k > 0$.

Now that we know what $\{C_i\}$ and $\{\xi_i\}$ must be, let's see if we can prove the formula for all $P \in \mathscr{P}_{2n-1}$. We know it is true for $P(x)$, a linear combination of elements of $\{x^k \pi(x)\}_{k=0}^{n-1}$ and $\{\pi(x)/(x - \xi_k)\}_{k=1}^{n}$. Let $P(x) \in \mathscr{P}_{2n-1}$. Then $P(x) = R(x)\pi(x) + R_1(x)$ where $R \in \mathscr{P}_{n-1}$ and $R_1 \in \mathscr{P}_{n-1}$. However, any polynomial $\varepsilon \mathscr{P}_{n-1}$ can be written as $\sum_{k=1}^{n} a_k (\pi(x)/(x - \xi_k))$ and in particular $R_1(x)$ can. Thus, since the quadrature formula holds for $R(x)\pi(x)$ and for $R_1(x)$, it hold for $P(x)$, which was an arbitrary polynomial $\varepsilon \mathscr{P}_{2n-1}$. ∎

(2) Jackson's Theorem for $C[0, 1]$

THEOREM (JACKSON) 1. $E_n(\mathscr{S}) \leq C/n$ for some constant C.

Proof. We will prove that for $f \in \mathscr{S}$, we can find a $P \in \mathscr{P}_n$ such that $|g(x) - P(x)| \leq C/n$ for all $x \in [1/4, 3/4]$ and for some C (which is independent of f and n), where $g(x)$ is defined on $[1/4, 3/4]$ by $g(x) = f(2x - 1/2)$. Then, if we set $t = 2x - 1/2$, we will get

$$\left| f(t) - P\left(\frac{t}{2} + \frac{1}{4}\right) \right| \leq \frac{C}{n}$$

for all $t \in [0, 1]$. Since $P(t/2 + 1/4)$ is a polynomial of degree at most n in t, we will have proven the theorem.

We extend $g(x)$ to $[0, 1]$ by defining $g(x) = g(1/4)$ for $x \in [0, 1/4]$ and $g(x) = g(3/4)$ for $x \in [3/4, 1]$. We observe that $|g(x) - g(y)| \leq 2|x - y|$ for $x, y \in [0, 1]$. This follows from the fact that $f \in \mathscr{S}$. Since

$$\max_{x \in [0,1]} g(x) - \min_{x \in [0,1]} g(x) \leq 2$$

and since we always have constants to approximate with, we can assume $|g(x)| \leq 1$ for $x \in [0, 1]$.

Let $\{\varphi_n(x)\}$ be a sequence of polynomials such that $\varphi_n(x) \varepsilon \mathscr{P}_n$, is non-negative and $\int_{-1}^{1} \varphi_n(x) dx = 1$. We observe that $\int_0^1 g(t)\varphi_n(x - t)dt$ is a polynomial in x of degree at most n.

$$\left| g(x) - \int_0^1 g(t)\varphi_n(x - t)dt \right|$$

$$\leq \left| g(x)\left(1 - \int_0^1 \varphi_n(x - t)dt\right) \right| + \int_0^1 |g(x) - g(t)|\varphi_n(x - t)dt$$

$$\leq 1 - \int_0^1 \varphi_n(x - t)dt + 2\int_0^1 |x - t|\varphi_n(x - t)dt$$

$$= \int_{-1}^1 \varphi_n(t)dt - \int_{x-1}^x \varphi_n(t)dt + 2\int_0^1 |x - t|\varphi_n(x - t)dt$$

$$= \int_{-1}^{x-1} \varphi_n(t)dt + \int_x^1 \varphi_n(t)dt + 2\int_{x-1}^x |t|\varphi_n(t)dt$$

$$\leq \int_{-1}^{-1/4} \varphi_n(t)dt + \int_{1/4}^{1} \varphi_n(t)dt + 2\int_{-1}^{1} |t|\varphi_n(t)dt$$

$$\leq 4\int_{-1}^{-1/4} |t|\varphi_n(t)dt + 4\int_{1/4}^{1} |t|\varphi_n(t)dt + 2\int_{-1}^{1} |t|\varphi_n(t)dt$$

$$\leq 6\int_{-1}^{1} |t|\varphi_n(t)dt$$

$$\leq 6\sqrt{\int_{-1}^{1} t^2\varphi_n(t)dt}\sqrt{\int_{-1}^{1} \varphi_n(t)dt}$$

$$= 6\sqrt{\int_{-1}^{1} t^2\varphi_n(t)dt}.$$

This is what we want to minimize subject to our constraints on $\varphi_n(t)$. It clearly will suffice to prove that for $n = 4K$ this expression is $\leq C/n$ since, for $n = 4K + 1$, $4K + 2$, or $4K + 3$, we will take the same polynomial as for $n = 4K$. Assuming that $n = 4K$, then $t^2\varphi_n(t) \in \mathscr{P}_{4K+2}$ and we use the quadrature formula with $2K + 2$ points. It is well known that the Legendre polynomials of even degree n are not zero at 0 and their zeroes are symmetric about the origin and the sequence of the zeroes of smallest absolute value approaches zero like π/n (for reference, see [32] p. 121).

We will denote the zeroes of the Legendre polynomial of degree $2K + 2$ by $\{\xi_i\}_{i=1}^{2K+2}$ and we observe that, since they are symmetric about the origin, ξ_{K+1} and ξ_{K+2} have the smallest absolute value. Then

$$\int_{-1}^{1} t^2\varphi_{4K}(t)dt = \sum_{i=1}^{2K+2} C_i\xi_i^2\varphi_{4K}(\xi_i)$$

$$\geq \xi_{K+2}^2 \sum_{i=1}^{2K+2} C_i\varphi_{4K}(\xi_i)$$

$$= \xi_{K+2}^2 \int_{-1}^{1} \varphi_{4K}(t)dt = \xi_{K+2}^2.$$

Thus, for a polynomial $\varphi_{4K}(t)$ satisfying our conditions,

$$\int_{-1}^{1} t^2\varphi_{4K}(t)dt \geq \xi_{K+2}^2.$$

Now let $\varphi(x) = c \prod_{i=1, i \neq K+1, K+2}^{2K+2} (x - \xi_i)^2$ where c is such that $\int_{-1}^{1} \varphi(x) = 1$. Then

$$\int_{-1}^{1} t^2\varphi(t)dt = C_{K+1}\xi_{K+1}^2\varphi(\xi_{K+1}) + C_{K+2}\xi_{K+2}^2\varphi(\xi_{K+2})$$

$$= \xi_{K+2}^2(C_{K+1}\varphi(\xi_{K+1}) + C_{K+2}\varphi(\xi_{K+2}))$$

$$= \xi_{K+2}^2 \int_{-1}^{1} \varphi(t)dt = \xi_{K+2}^2.$$

Thus the minimum of $\int_{-1}^{1} t^2 \varphi_{4K}(t)dt$ is ξ_{K+2}^2 and we always pick $\varphi_{4K}(t)$ so as to achieve this minimum. Since, for large K, $\xi_{K+2} \le 4/4K$ we can get $6\sqrt{\int_{-1}^{1} t^2 \varphi_{4K}(t)dt} = 6\xi_{K+2} \le 6/K$ which is $24/n$ for $n = 4K$. ∎

The next theorem shows why we are so concerned about $E_n(\mathscr{S})$. We show that knowing $E_n(\mathscr{S})$ enables us to estimate $E_n(f)$ for any f.

THEOREM (JACKSON) 2. *Let* $f \in C[0,1]$. *Then* $E_n(f) \le C_0 \omega(f, 1/n)$ *where* $C_0 = 2 + C$ *and* C *is from the previous theorem.*

Proof. $L(x)$ be a function such that $L(k/n) = f(k/n)$ $k = 0, \ldots, n$ and linear in between. Let $x \in [k/n, (k+1)/n]$. Then

$$|f(x) - L(x)| \le \left|f(x) - f\left(\frac{k}{n}\right)\right| + \left|f\left(\frac{k}{n}\right) - L\left(\frac{k}{n}\right)\right|$$
$$+ \left|L\left(\frac{k}{n}\right) - L(x)\right|$$
$$\le \omega\left(f, \frac{1}{n}\right) + 0 + \omega\left(f, \frac{1}{n}\right) = 2\omega\left(f, \frac{1}{n}\right).$$

The largest slope of $L(x)$ is the largest of $n|f((k+1)/n) - f(k/n)|$ which is $\le n\omega(f, 1/n)$. Therefore,

$$\frac{L(x)}{n\omega\left(f, \frac{1}{n}\right)} \in \mathscr{S}.$$

By the previous theorem, there is a $P \in \mathscr{S}_n$ such that

$$\left\|\frac{L(x)}{n\omega\left(f, \frac{1}{n}\right)} - P(x)\right\| \le \frac{C}{n}.$$

Letting $Q_n(x) = n\omega(f, 1/n) P(x) \in \mathscr{S}_n$, we have

$$\|L(x) - Q_n(x)\| \le C\omega(f, 1/n).$$

Therefore

$$E_n(f) \le \|f(x) - Q_n(x)\| \le \|f(x) - L(x)\| + \|L(x) - Q_n(x)\|$$
$$\le 2\omega\left(f, \frac{1}{n}\right) + C\omega\left(f, \frac{1}{n}\right) = (2 + C)\omega\left(f, \frac{1}{n}\right). \ \blacksquare$$

Corollary. If $f \in \text{lip } \alpha$, then $E_n(f) \le C_1/n^\alpha$ for some C_1 and all n.

This theorem is the best possible in the sense that we might be able to get a constant smaller than $(2 + C)$ but we could not replace $\omega(f, 1/n)$

by any other function of f and n which, for all f, would go to zero faster, as n approached infinity. We could not even do it for the class $\mathscr{S}$. This is seen by taking the function $f(x)$ which is $(-1)^k/2n$ at k/n and linear in between. As we have seen, for this $f(x)$, $f \in \mathscr{S}$ and $E_n(f) = 1/2n = 1/2\, \omega(f, 1/n)$.

This is not to say that $E_n(f)$ never goes to zero faster than $\omega(f, 1/n)$. First of all, if $f \in \mathscr{P}_n$, then $E_n(f) = E_{n+1}(f) = \cdots = 0$ and yet $\omega(f, 1/n) \neq 0$ (unless f is constant). Moreover, for the class $C^p[0, 1]$ of functions on $[0, 1]$ with p continuous derivatives, we can do better, as we shall see.

Corollary. If $f \in C^1[0, 1]$, then $E_n(f) \leq (C_0/n)\|f'\|$.

THEOREM 3. Let $f \in C^1[0, 1]$. Then $E_n(f) \leq (C_0/n) E_{n-1}(f')$.

Proof. Let $P_{n-1} \in \mathscr{P}_{n-1}$ be such that $\|f' - P_{n-1}\| = E_{n-1}(f')$ and let $P_n(x) = \int_0^x P_{n-1}(t)dt$. Then, if $g(x) = f(x) - P_n(x)$,

$$g'(x) = f'(x) - P_{n-1}(x)$$

and $\|g'(x)\| = E_{n-1}(f')$. By the previous corollary,

$$E_n(g) \leq (C_0/n) E_{n-1}(f').$$

But

$$E_n(g) = \min_{P \in \mathscr{P}_n} \|g - P\| = \min_{P \in \mathscr{P}_n} \|f - P_n - P\|$$
$$= \min_{P \in \mathscr{P}_n} \|f - P\| = E_n(f).$$

Thus $E_n(f) \leq (C_0/n) E_{n-1}(f')$. ∎

LEMMA 4. Let p and n be positive integers with $n \geq p + 1$ and let $\omega(x)$ be a modulus of continuity. Then $\omega(1/(n - p)) \leq p(p + 1)\, \omega(1/n)$.

Proof. Since $1/(n - p) \leq p(p + 1)/n$, we have

$$\omega\!\left(\frac{1}{n-p}\right) \leq \omega\!\left(\frac{p(p+1)}{n}\right) = \omega\!\left(\frac{1}{n} + \cdots + \frac{1}{n}\right)$$
$$\leq p(p+1)\omega\!\left(\frac{1}{n}\right). \quad\blacksquare$$

THEOREM (JACKSON) 5. Let $f \in C^p[0, 1]$. Then, if $n \geq p+1$,

$$E_n(f) \leq (C_p/n^p)\omega(f^{(p)}, 1/n)$$

where C_p is a constant which depends only on p.

Proof.

$$E_n(f) \le \frac{C_0}{n} E_{n-1}(f') \le \frac{C_0^2}{n(n-1)} E_{n-2}(f'')$$

$$\le \frac{C_0^p}{n(n-1) \cdots (n-p+1)} E_{n-p}(f^{(p)})$$

$$\le \frac{C_0^{p+1}}{n(n-1) \cdots n-p+1} \omega\left(f^{(p)}, \frac{1}{n-p}\right)$$

$$\le \frac{C_0^{p+1} p(p+1)}{n(n-1) \cdots (n-p+1)} \omega\left(f^{(p)}, \frac{1}{n}\right)$$

$$\le \frac{C_p}{n^p} \omega\left(f^{(p)}, \frac{1}{n}\right)$$

where

$$C_p = C_0^{p+1} p(p+1) \max_{n \ge p+1} \left[\frac{n^p}{n(n-1) \cdots (n-p+1)}\right] \quad \blacksquare$$

Corollary. If $f \in C^p[0, 1]$ and $f^{(p)} \in \text{lip } \alpha$, $0 < \alpha \le 1$, then

$$E_n(f) \le \frac{C_p}{n^{p+\alpha}} \quad \text{for } n \ge p+1.$$

(3) Jackson's Theorem for $C^*[-\pi, \pi]$

Let $f \in C^*[-\pi, \pi]$ and consider $S_n(f, x)$, the operator which assigns to each f, its n-th partial sum. $S_n(f, x)$ can be written as the convolution of $f(x)$ and the Dirichlet kernel,

$$\frac{1}{2\pi} \frac{\sin\left(n + \frac{1}{2}\right)x}{\sin \frac{1}{2} x}.$$

That is,

$$S_n(f, x) = \frac{1}{2\pi} \int_{-\pi}^{\pi} f(\theta) \frac{\sin\left(n + \frac{1}{2}\right)(x - \theta)}{\sin \frac{1}{2}(x - \theta)} d\theta.$$

As we shall see later, $S_n(f, x)$ is a very poor choice of trigonometric polynomial to approximate $f(x)$, as there are functions for which $S_n(f, x)$ diverges as n approaches infinity.

We next consider the Cesàro operator $\sigma_n(f, x)$. This can be written as the convolution of $f(x)$ and the Fejér kernel

$$\frac{1}{2\pi n} \frac{\sin^2\left(\frac{n}{2} + \frac{1}{2}\right)x}{\sin^2 \frac{x}{2}}.$$

That is,

$$\sigma_n(f, x) = \frac{1}{2\pi n} \int_{-\pi}^{\pi} f(\theta) \frac{\sin^2\left(\frac{n+1}{2}\right)(x-\theta)}{\sin^2\left(\frac{x-\theta}{2}\right)} d\theta.$$

$\sigma_n(f, x)$ is a better choice for approximating $f(x)$ than $S_n(f, x)$, since, for every $f \in C^*[-\pi, \pi]$, $\sigma_n(f, x)$ converges uniformly to $f(x)$. However, we are interested in how closely we can approximate $f(x)$ by n-th degree trigonometric polynomials and, as we shall see, the Cesàro sums are not always as close as we can get with other polynomials.

We notice that, in some sense, the Fejér kernel can be considered as the square of the Dirichlet kernel. (The only two changes are $n/2 + 1/2$ instead of $n + 1/2$, which is needed to get an n-th degree polynomial, and $1/2\pi n$ instead of $1/2\pi$ which is necessary to have the integral of the kernel equal to 1.) Jackson's idea was to take the square of the Fejér kernel, with the same two modifications. We first consider

$$\left(\frac{\sin\left(\frac{n+1}{2}\right)x}{\sin\frac{x}{2}}\right)^4$$

and observe that since the Fejér kernel is a trigonometric polynomial of degree n, what we have is of degree $2n$. We thus have to replace n by $[n/2]$, the greatest integer in $n/2$. We then define

$$K_n(x) = \frac{1}{C_n} \left(\frac{\sin\left(\left[\frac{n}{2}\right] + 1\right)\left(\frac{x}{2}\right)}{\sin\frac{x}{2}}\right)^4$$

where C_n is such that $\int_{-\pi}^{\pi} K_n(x)dx = 1$. The Jackson operator $J_n(f, x)$ is now defined as the convolution of $f(x)$ and $K_n(x)$. That is

$$J_n(f, x) = \int_{-\pi}^{\pi} f(\theta) K_n(x - \theta) d\theta.$$

LEMMA 6.
$$C_n = \frac{2\pi}{3}\left(\left[\frac{n}{2}\right] + 1\right)\left(2\left(\left[\frac{n}{2}\right] + 1\right)^2 + 1\right).$$

Proof.
$$C_n = \int_{-\pi}^{\pi} \left(\frac{\sin\left(\left[\frac{n}{2}\right] + 1\right)\left(\frac{x}{2}\right)}{\sin\frac{x}{2}}\right)^4 dx$$

and, for simplicity, we set $m = [n/2] + 1$. From our knowledge of the Fejér kernel, we know that

$$\left(\frac{\sin\frac{mt}{2}}{\sin\frac{t}{2}}\right)^2 = m + 2[(m-1)\cos t + (m-2)\cos 2t$$
$$+ \cdots + \cos(m-1)t].$$

Since $\int_{-\pi}^{\pi} \cos jt \cos kt\, dt = \pi \delta_{jk}$, we easily get that

$$C_n = 2\pi m^2 + 4\pi[(m-1)^2 + (m-2)^2 + \cdots + 1^2].$$

The lemma now follows by induction. ∎

LEMMA 7. $\omega(f, \delta) \leq (n\delta + 1)\omega(f, 1/n)$.

Proof.
$$\omega(f, \delta) = \omega\left(f, \frac{n\delta}{n}\right)$$
$$\leq \omega\left(f, \frac{[n\delta] + 1}{n}\right)$$
$$\leq ([n\delta] + 1)\omega\left(f, \frac{1}{n}\right)$$
$$\leq (n\delta + 1)\omega\left(f, \frac{1}{n}\right) \quad \blacksquare$$

THEOREM (JACKSON) 8. Let $f \in C^*[-\pi, \pi]$. Then
$$\|f(x) - J_n(f, x)\| \leq 17\omega(f, 1/n).$$

Proof. Let $x \in [-\pi, \pi]$. Then,

$$|f(x) - J_n(f, x)| = \left| f(x) - \int_{-\pi}^{\pi} f(\theta) K_n(x - \theta) d\theta \right|$$

$$= \left| f(x) - \int_{-\pi}^{\pi} f(x - \theta) K_n(\theta) d\theta \right|$$

$$= \left| \int_{-\pi}^{\pi} (f(x) - f(x - \theta)) K_n(\theta) d\theta \right|$$

$$\leq \int_{-\pi}^{\pi} |f(x) - f(x - \theta)| K_n(\theta) d\theta$$

$$\leq \int_{-\pi}^{\pi} \omega(f, |\theta|) K_n(\theta) d\theta$$

$$\leq \int_{-\pi}^{\pi} (n|\theta| + 1) \omega\left(f, \frac{1}{n}\right) K_n(\theta) d\theta$$

$$= 2n\omega\left(f, \frac{1}{n}\right) \int_0^{\pi} \theta K_n(\theta) d\theta + \omega\left(f, \frac{1}{n}\right).$$

Thus, to prove the theorem, we just have to prove that

$$\int_0^{\pi} \theta K_n(\theta) d\theta \leq \frac{8}{n}.$$

$$\int_0^{\pi} \theta K_n(\theta) d\theta = 4 \int_0^{\pi/2} \theta K_n(2\theta) d\theta$$

$$= \frac{4}{C_n} \int_0^{1/([n/2]+1)} \theta \left(\frac{\sin\left(\left[\frac{n}{2}\right] + 1\right)\theta}{\sin \theta} \right)^4 d\theta$$

$$+ \frac{4}{C_n} \int_{1/([n/2]+1)}^{\pi/2} \theta \left(\frac{\sin\left(\left[\frac{n}{2}\right] + 1\right)\theta}{\sin \theta} \right)^4 d\theta$$

$$\leq \frac{4}{C_n} \int_0^{1/([n/2]+1)} \theta \left(\left[\frac{n}{2}\right] + 1\right)^4 d\theta + \frac{4}{C_n} \int_{1/([n/2]+1)}^{\infty} \theta \left(\frac{\pi}{2\theta}\right)^4 d\theta$$

$$= \frac{2}{C_n} \left(\left[\frac{n}{2}\right] + 1\right)^2 + \frac{\pi^4}{8 C_n} \left(\left[\frac{n}{2}\right] + 1\right)^2$$

$$= \frac{\left(2 + \frac{\pi^4}{8}\right)\left(\left[\frac{n}{2}\right] + 1\right)}{\frac{2\pi}{3}\left(2\left(\left[\frac{n}{2}\right] + 1\right)^2 + 1\right)}$$

$$< \frac{4}{\left[\frac{n}{2}\right] + 1} < \frac{8}{n}. \blacksquare$$

We note that this is essentially the best we can do as we have seen an $f \in \mathscr{S}^*$ such that $E_n^*(f) \geq 1/(2(n+1))$ which is $\omega(f, 1/(2(n+1)))$.

Corollary. Let f be in $C^*[-\pi, \pi]$. Then $E_n^*(f) \leq 17\omega(f, 1/n)$.

Corollary. $E_n^*(\mathscr{S}^*) \leq 17/n$.

Corollary. If f is in (lip α) $\cap C^*[-\pi, \pi]$, where $0 < \alpha < 1$, then $E_n^*(f) \leq C/n^\alpha$ for some C and all n.

Definition. $C^{P*}[-\pi, \pi] = \{f \in C^*[-\pi, \pi]: f', f'', f''', \ldots f^{(P)}$ exist and are in $C^*[-\pi, \pi]$.$\}$

Corollary. If $f \in C^{1*}[-\pi, \pi]$, then $E_n^*(f) \leq (17/n)\|f'\|$.

As we saw with the previous Jackson theorem, this Jackson theorem gives us a basically best possible upper bound for $E_n^*(f)$. That is, we could never replace $17\omega(f, 1/n)$ by anything smaller except possibly $c\omega(f, 1/n)$ (where $c < 17$), and yet have theorem true for all $f \in C^*[-\pi, \pi]$. This is seen by considering the example used to prove $E_n^*(\mathscr{S}^*) \geq \pi/(2(n+1))$.

Actually, we could have gotten the previous Jackson theorem (about algebraic polynomials) from this one. Let $f \in C[-1, 1]$ and define $g(\theta) = f(\cos\theta)$. Then $g \in C^*[-\pi, \pi]$ and g is even. We know we can find a trigonometric polynomial $\varepsilon \mathscr{T}_n$, $T_n(\theta)$, such that

$$|g(\theta) - T_n(\theta)| \leq 17\omega\left(g, \frac{1}{n}\right).$$

Without loss of generality, since $g(\theta)$ is even, we can assume that $T_n(\theta)$ is, also. That is, $T_n(\theta)$ involves only cosines. Now let $\theta = \arccos x$. Then $T_n(\arccos x)$ is a polynomial of degree n in x and $g(\arccos x) \equiv f(x)$. Thus $|f(x) - T_n(\arccos x)| \leq 17\omega(g, 1/n)$. Our theorem will then be proven if we can show $\omega(g, 1/n) \leq \omega(f, 1/n)$. We observe that if $x_1 = \cos\theta$, and $x_2 = \cos\theta_2$, then $|x_1 - x_2| = |\cos\theta_1 - \cos\theta_2| \leq |\theta_1 - \theta_2|$. Thus

$$\omega(g, \delta) = \max_{|\theta_1 - \theta_2| \leq \delta} |g(\theta_1) - g(\theta_2)| = \max_{|\theta_1 - \theta_2| \leq \delta} |f(\cos\theta_1) - f(\cos\theta_2)|$$

$$\leq \max_{|x_1 - x_2| \leq \delta} |f(x_1) - f(x_2)| = \omega(f, \delta).$$

We will now show, that, as before for the class $C^P[0, 1]$, for the class $C^{P*}[-\pi, \pi]$ we can improve on Jackson's theorems. First we need the following

LEMMA 9. If $f \in L^1[-\pi, \pi]$, $\int_{-\pi}^{\pi} f(x)dx = 0$ and

$$\|f(x) - \sum_{n=0}^{N}(a_n \cos nx + b_n \sin nx)\| = \delta,$$

then $|a_0| \leq \delta$.

Proof.

$$2\pi |a_0| = |\int_{-\pi}^{\pi}[f(x) - \sum_{n=0}^{N}(a_n \cos nx + b_n \sin nx)]dx|$$

$$\leq \int_{-\pi}^{\pi}|f(x) - \sum_{n=0}^{N}(a_n \cos nx + b_n \sin nx)|dx \leq 2\pi\delta.$$

THEOREM 10. *If* $f \in C^{1*}[-\pi,\pi]$, *then* $E_n^*(f) \leq (34/n)E_n^*(f')$.

Proof. Let $T_n(x)$ be the best n-th degree approximant to $f'(x)$. Let

$$\bar{T}_n(x) = T_n(x) - \frac{1}{2\pi}\int_{-\pi}^{\pi} T_n(x)dx.$$

By lemma 9, since $\int_{-\pi}^{\pi} f'(t)dx = 0$,

$$\left|\frac{1}{2\pi}\int_{-\pi}^{\pi} T_n(x)dx\right| \leq \|f'(x) - T_n(x)\| = E_n^*(f').$$

Now let $g(x) = f(x) - \int_0^x \bar{T}_n(t)dt$. Then

$$\|g'(x)\| = \|f'(x) - \bar{T}_n(x)\|$$

$$\leq \|f'(x) - T_n(x)\| + \left\|\frac{1}{2\pi}\int_{-\pi}^{\pi} T_n(x)dx\right\| \leq 2E_n^*(f')$$

Then, by the fourth corollary to theorem 8,

$$E_n^*(g) \leq \frac{17}{n}\|g'\| \leq \frac{34}{n}E_n^*(f').$$

Since $f(x)$ differs from $g(x)$ by a trigonometric polynomial of degree n, $E_n^*(f) = E_n^*(g)$ and thus

$$E_n^*(f) \leq \frac{34}{n} E_n^*(f') \quad \blacksquare$$

THEOREM 11. *If* $f \in C^{P*}[-\pi,\pi]$, *then*

$$E_n^*(f) \leq \frac{17 \cdot 34^P}{n^P} \omega\left(f^{(P)}, \frac{1}{n}\right)$$

Proof.

$$E_n^*(f) \leq \frac{34}{n} E_n^*(f') \leq \frac{34^2}{n^2} E_n^*(f'') \leq \cdots$$

$$\leq \frac{34^P}{n^P} E_n^*(f^{(P)}) \leq \frac{17 \cdot 34^P}{n^P} \omega\left(f^{(P)}, \frac{1}{n}\right). \blacksquare$$

Corollary. If $f \in C^{P*}[-\pi, \pi]$, and $f^{(P)} \in \text{lip } \alpha$, then

$$E_n^*(f) \leq \frac{17 \cdot 34^P M}{n^{P+\alpha}}.$$

(4) Favard's Theorem

We have previously derived an upper bound for $E_n^*(\mathscr{S}^*)$ of $(17/n)$ and a lower bound of $(\pi/((n+1))$. We derived the upper bound by taking, for any $f \in \mathscr{S}^*$, the convolution of f and the Jackson kernel, $K_n(x)$, and using that to approximate $f(x)$. $K_n(x)$ had the following properties:

(a) $K_n(x) \in \mathscr{T}_n$
(b) $K_n(x) \geq 0$
(c) $\int_{-\pi}^{\pi} K_n(x) dx = 1$
(d) $\int_0^{\pi} x K_n(x) \leq 8/n$.

The idea of Favard was to take a kernel $F_n(x)$ with the following properties:

(a) $F_n(x) \in \mathscr{T}_n$
(b) $F_n(x)$ is even in x
(c) $\int_{-\pi}^{\pi} F_n(x) dx = 1$
(d) $\int_{((j\pi)/(n+1))}^{((j+1)\pi/(n+1))} F_n(x) dx = 0$ for $j = 1, 2, 3, \ldots, n$.

We will first prove the existence of such a kernel. From the first two conditions we know that $F_n(x) = a_0 + a_1 \cos x + \ldots + a_n \cos nx$. The third condition tells us that $a_0 = 1/2\pi$. The last condition is equivalent to

$$\int_0^{((j\pi)/(n+1))} F_n(x) dx = \frac{1}{2} \text{ for } j = 1, 2, \ldots, n+1$$

That condition, for $j = n+1$, is obvious from properties (b) and (c). For $j = 1, 2, \ldots, n$, we get

$$\frac{1}{2\pi} \frac{j\pi}{n+1} + a_1 \sin\left(\frac{j\pi}{n+1}\right) + \frac{a_2}{2} \sin\left(\frac{2j\pi}{n+1}\right) + \cdots$$
$$+ \frac{a_n}{n} \sin\left(\frac{nj\pi}{n+1}\right) = \frac{1}{2}$$

or

$$a_1 \sin\left(\frac{j\pi}{n+1}\right) + \frac{a_2}{2} \sin\left(\frac{2j\pi}{n+1}\right) + \cdots + \frac{a_n}{n} \sin\left(\frac{nj\pi}{n+1}\right)$$

$$= \frac{n+1-j}{2(n+1)}$$

What we therefore want is that the $n \times n$ matrix whose $j - k$ term is $1/k \sin(kj\pi/(n+1))$ should be nonsingular. If, however, it were singular, it would map a non-zero vector $(b_1, b_2, \ldots, b_n)$ to $(0, 0, \ldots, 0)$. Then $b_1 \sin x + b_2/2 \sin(2x) + \cdots + b_n/n \sin(nx)$ would vanish at $x = j\pi/(n+1)$ for $j = 0, \pm 1, \pm 2, \ldots, \pm(n+1)$ which is impossible for a polynomial $\varepsilon \mathcal{T}_n$.

Actually, if we wanted to work harder, we could prove that

$$\frac{1}{2\pi}\left[1 + \sum_{j=1}^{n} \frac{j\pi}{2(n+1)} \cot \frac{j\pi}{2(n+1)} \cos jx\right]$$

satisfies our requirements, but the knowing of the existence of such a kernel is enough for us.

Let $H_n(x) = \int_0^x F_n(t) dt - 1/2$ for $x \in (0, \pi)$ and for $x \in (-\pi, 0)$, let $H_n(x) = -H_n(-x)$. Thus $H'_n(x) = F_n(x)$ for $x \in (-\pi, 0) \cup (0, \pi)$. Also, $H_n(j\pi/(n+1)) = 0$ for $j = \pm 1, \pm 2, \ldots, \pm(n+1)$ and, moreover, those are the only zeroes of $H_n(x) \in (-\pi, 0) \cup (0, \pi)$. (The last statement is true because otherwise Rolle's theorem would give us at least $2n+1$ zeroes for $F_n(x)$ in $(-\pi, \pi)$, which is impossible since $F_n(x) \in \mathcal{T}_n$.) Thus, $H_n(x) < 0$ for $x \in (0, \pi/(n+1))$, $H_n(x) > 0$ for $x \in (\pi/(n+1), 2\pi/(n+1))$, etc.

We would now like to estimate the difference between $f(x)$ and $f * F_n(x)$ for any $f \in \mathcal{S}^*$. For that, though, we need to know the following easily proven alternative definition of the class $\mathcal{S}^*$.

THEOREM 12. $f \in \mathcal{S}^*$ iff $f(-\pi) = f(\pi)$, f is absolutely continuous and $|f'(x)| \le 1$ a.e.

Now, let f be in $\mathcal{S}^*$. Then

$$\max_x \left| f(x) - \int_{-\pi}^{\pi} f(t-x) F_n(t) dt \right|$$

$$= \max_x \left| \int_{-\pi}^{\pi} (f(x) - f(t-x)) F_n(t) dt \right|$$

$$= \max_x \left| \int_{-\pi}^{\pi} f'(t-x) H_n(t) dt \right|$$

$$\le \int_{-\pi}^{\pi} |H_n(t)| dt$$

$$= \int_{-\pi}^{\pi} S_n(t) H_n(t) dt$$

where $S_n(t)$ a "saw tooth" function, is 1 where $H_n(t)$ is positive and -1

where it is negative.

We notice that if for some x, $f'(t - x) = S_n(t)$ then we would have equality, Let us therefore pick out $x = 0$ and define a function

$$f_n(t) = \int_0^t S_n(y)\,dy + \frac{\pi}{2(n+1)}.$$

Then $f_n(x)$ would be the function which is

$$(-1)^k \frac{\pi}{2(n+1)} \quad \text{at } x = \frac{k\pi}{n+1}$$

and linear in between. It is easily seen that $f_n \in \mathscr{S}^*$. Moreover,

$$\max_x \left| f_n(x) - \int_{-\pi}^{\pi} f_n(t-x) F_n(t)\,dt \right|$$

$$= \left| f_n(0) - \int_{-\pi}^{\pi} f_n(t) F_n(t)\,dt \right|$$

$$= \int_{-\pi}^{\pi} S_n(t) H_n(t)\,dt.$$

Since $f_n(x + (2\pi/(n+1))) \equiv f_n(x)$, $f_n(x)$ is orthogonal to all of $\mathscr{T}_n$ and in particular to $F_n(x)$. Thus

$$\int_{-\pi}^{\pi} S_n(t) H_n(t)\,dt = |f_n(0)| = \frac{\pi}{2(n+1)}.$$

Since, for any $f \in \mathscr{S}^*$,

$$E_n^*(f) \leq \max_x \left| f(x) - \int_{-\pi}^{\pi} f(t-x) F_n(t)\,dt \right|$$

$$\leq \int_{-\pi}^{\pi} S_n(t) H_n(t)\,dt = \frac{\pi}{2(n+1)}$$

we have proven that $E_n^*(\mathscr{S}^*) \leq \pi/2(n+1)$, which is exactly the lower bound gotten before (by proving $E_n^*(f_n) \geq \pi/2(n+1)$). We have thus proven the following:

THEOREM (FAVARD) 13. $E_n^*(\mathscr{S}^*) = \pi/2(n+1)$.

CHAPTER V

INVERSE THEOREMS FOR PERIODIC FUNCTIONS

In previous chapters, we concerned ourselves with questions of the type "If $f(x)$ has such and such properties or if $f(x)$ is in such and such a class of functions, how closely do polynomials approximate $f(x)$?"

In this chapter, we concern ourselves with the opposite kinds of questions. We ask "If $f(x)$ can be approximated by polynomials to within a certain degree of closeness, what properties does $f(x)$ have (or to what class of functions does $f(x)$ belong)?"

We will concern ourselves with approximation only by trigonometric polynomials since the answers for algebraic polynomials are not as complete as with trigonometric polynomials. Accordingly, in this chapter, we will always assume that $f(x)$ is periodic with period 2π.

$C_1, C_2, \ldots$ will denote absolute constants; *i.e.* constants independent of any of the variables including N and h. We will not strive for the best constants, but will be content with any constants. If C_1 is used in one theorem, it will not be the same (numerically) as the C_1 used in a different theorem.

(1) Bernstein's Theorems

Jackson's Theorem says that if $0 < \alpha \leq 1$ and $f \in \text{lip } \alpha$, then $E_N{}^*(f) \leq C_1/N^\alpha$. Bernstein proved the following inverse

THEOREM 1. (a) Let $0 < \alpha < 1$. If $E_N{}^*(f) \leq C/N^\alpha$ then $f \in \text{lip } \alpha$.
(b) If $E_N{}^*(f) \leq C/N$ then $\omega(f, h) \leq C_2 h \log(1/h)$ (for small, positive h).

Proof. Let $0 < \alpha \leq 1$ and assume there is a $T_N \in \mathcal{T}_N$ for each N, such that $||f - T_N|| \leq C/N^\alpha$.

Let
$$Q_N(x) = T_{2^N}(x) - T_{2^{N-1}}(x)$$
for $N = 1, 2, \ldots$, and let $Q_0(x) = T_1(x)$. Then
$$\|Q_N\| \leq \|T_{2^N} - f\| + \|f - T_{2^{N-1}}\|$$
$$\leq \frac{C}{2^{N\alpha}} + \frac{C}{2^{(N-1)\alpha}} = \frac{C_1}{2^{N\alpha}}.$$

Actually, the methods used in deriving this inequality are not applicable in the case $N = 0$, but the overall inequality is. Therefore,
$$\|Q_N'\| \leq 2^N \|Q_N\| \leq C_1 \, 2^{N(1-\alpha)}.$$

If we set
$$S_N(x) = \sum_{K=0}^{N} Q_K(x)$$
we see that
$$S_N(x) \equiv T_{2^N}(x),$$
and since $\|f - T_{2^N}\|$ goes to 0 as N approaches ∞ we have that
$$f(x) \equiv \sum_{K=0}^{\infty} Q_K(x).$$

Let $0 < h < 1$ and let n be a positive integer which we'll determine later. Then
$$|f(x+h) - f(x)| = \left|\sum_{K=0}^{\infty} (Q_K(x+h) - Q_K(x))\right|$$
$$\leq \left|\sum_{K=n}^{n-1} (Q_K(x+h) - Q_K(x))\right|$$
$$+ \left|\sum_{K=n}^{\infty} (Q_K(x+h) - Q_K(x))\right|$$
$$\leq h \sum_{K=0}^{n-1} \|Q_K'\| + 2 \sum_{K=n}^{\infty} \|Q_K\|$$
$$\leq h\, C_1 \sum_{K=0}^{n-1} 2^{K(1-\alpha)} + 2C_1 \sum_{K=n}^{\infty} 2^{-K\alpha}.$$

(a) If $0 < \alpha < 1$, we get
$$|f(x+h) - f(x)| \leq h\, C_1 \left(\frac{2^{n(1-\alpha)} - 1}{2^{1-\alpha} - 1}\right) + \frac{2C_1}{2^{n\alpha}} \frac{1}{1 - 2^{-\alpha}}$$
$$\leq C_2(h 2^{n(1-\alpha)} + 2^{-n\alpha}).$$

We let $n = [1 - \log_2 h]$ and then $1/h \leq 2^n \leq 2/h$. Thus,

$$|f(x + h) - f(x)| \leq C_2(h\left(\frac{2}{h}\right)^{1-\alpha} + h^\alpha)$$

$$= C_3 h^\alpha$$

and $f \in \text{lip } \alpha$.

(b) If $\alpha = 1$, we get

$$|f(x + h) - f(x)| \leq C_1 h n + \frac{2C_1}{2^n} \frac{1}{1 - \frac{1}{2}}$$

$$= C_1(h n + 4 \cdot 2^{-n})$$

We now let $n = [\log_2(1/h)]$ and then $2^n \leq 1/h < 2^{n+1}$. Thus

$$|f(x + h) - f(x)| \leq C_1(h \log_2(1/h) + 8h)$$

$$\leq C_2 h \log\left(\frac{1}{h}\right)$$

for small, positive h.

Among the corollaries to Jackson's theorem was that if $f \in C^{P*}[-\pi, \pi]$ and $f^{(P)} \in \text{lip } \alpha$ $(0 < \alpha \leq 1)$ then $E_N^*(f) \leq C/N^{P+\alpha}$. For this we also have an inverse theorem due to Bernstein.

THEOREM 2. Let P be any positive integer.

(a) If $0 < \alpha < 1$ and $E_N^*(f) \leq C/N^{P+\alpha}$ then $f \in C^{P*}[-\pi, \pi]$ and $f^{(P)} \in \text{lip } \alpha$.

(b) If $E_N^*(f) \leq C/N^{P+1}$ then $f \in C^{P*}[-\pi, \pi]$ and

$$\omega(f^{(P)}, h) \leq C_4 h \log 1/h \quad \text{(for small, positive } h\text{)}.$$

Proof. Let $0 < \alpha \leq 1$. As before, we write

$$f(x) = \sum_{K=0}^{\infty} Q_K(x)$$

and have

$$\|Q_K\| \leq \|T_{2^K} - f\| + \|f - T_{2^{K-1}}\| \leq \frac{C}{2^{K(P+\alpha)}} + \frac{C}{2^{(K-1)(P+\alpha)}}$$

$$= \frac{C_1}{2^{K(P+\alpha)}}.$$

Therefore, by repeated application of $\|Q_K'\| \leq 2^K \|Q_K\|$ we get

$$\|Q_K^{(P)}\| \leq (2^K)^P \|Q_K\| \leq \frac{C_1}{2^{K\alpha}}.$$

Thus, $\sum_{K=0}^{\infty} Q_K^{(P)}$ converges uniformly, and $f(x)$ is P times differentiable. Then,

$$E_{2N}^*(f^{(P)}) \leq \|f^{(P)} - \sum_{K=0}^{N} Q_K^{(P)}\| = \|\sum_{K=N+1}^{\infty} Q_K^{(P)}\|$$

$$\leq C_1 \sum_{K=N+1}^{\infty} \frac{1}{2^{K\alpha}} \leq \frac{C_2}{2^{N\alpha}}.$$

(a) For $0 < \alpha < 1$ and for any $K > 0$, choose N so that

$$2^N \leq K < 2^{N+1}.$$

Then,

$$E_K^*(f^{(P)}) \leq E_{2N}^*(f^{(P)}) \leq C_2(2^N)^{-\alpha} \leq C_2\left(\frac{K}{2}\right)^{-\alpha} = \frac{C_3}{K^\alpha}.$$

Thus, by the previous theorem, $f^{(P)} \in \text{lip } \alpha$.

(b) If $\alpha = 1$, then we get $E_{2N}^*(f^{(P)}) \leq C_2/2^N$. For any $K > 0$, choose N so that $2^N \leq K < 2^{N+1}$. Then $E_K^*(f^{(P)}) \leq E_{2N}^*(f^{(P)}) \leq C_2/2^N \leq 2C_2/K$. Again by the previous theorem, we get $\omega(f^{(P)}, h) \leq C_4 h \log(1/h)$ (for small positive h). ∎

Summary. (1) $0 < \alpha < 1$: $E_N^*(f) \leq C/N^\alpha$ iff $f \in \text{lip } \alpha$.
(2) $E_N^*(f) \leq C/N \longrightarrow \omega(f, \delta) \leq C_3 \delta \log(1/\delta)$.
(3) $\omega(f, \delta) \leq C \delta \longrightarrow E_N^*(f) \leq C/N$.

We would like to bring (2) and (3) closer together to get an iff condition for $E_N^*(f)$ to be $\leq C/N$, similar to the iff condition in (1). It is tempting to say that the implication in (3) should be reversed, but that is wrong as seen in the following

THEOREM 3. *There is an $f(x) \in C^*[-\pi, \pi]$ such that $E_N^*(f) \leq C/N$ but f is not in lip 1.*

Proof. Let

$$f(x) = \sum_{K=1}^{\infty} \frac{\sin Kx}{K^2}$$

Then

$$E_N^*(f) \leq \|f(x) - \sum_{K=1}^{N} \frac{\sin Kx}{K^2}\| = \|\sum_{K=N+1}^{\infty} \frac{\sin Kx}{K^2}\|$$

$$\leq \sum_{K=N+1}^{\infty} \frac{1}{K^2} < \frac{1}{N}.$$

To prove $f \notin \text{lip } 1$, it suffices to prove
$$\left| \frac{f(h) - f(0)}{h} \right|$$
is unbounded as h approaches 0^+.

$$\left| \frac{f(h) - f(0)}{h} \right| = \left| \sum_{n=1}^{\infty} \frac{\sin nh}{n^2 h} \right|$$

$$\geq \left| \sum_{n=1}^{[\pi/2h]-1} \frac{\sin nh}{n^2 h} \right| - \left| \sum_{n=[\pi/2h]}^{\infty} \frac{\sin nh}{n^2 h} \right|$$

$$\geq \sum_{n=1}^{[\pi/2h]-1} \frac{\frac{2nh}{\pi}}{n^2 h} - \frac{1}{h} \sum_{n=[\pi/2h]}^{\infty} \frac{1}{n^2}$$

$$\geq \frac{2}{\pi} \int_{1}^{[\pi/2h]} \frac{dx}{x} - \frac{2}{h} \int_{[\pi/2h]}^{\infty} \frac{dx}{x^2}$$

$$= \frac{2}{\pi} \log \left[\frac{\pi}{2h} \right] - \frac{2}{h} \frac{1}{\left[\frac{\pi}{2h} \right]}$$

which is unbounded as h approaches 0^+. ∎

Note. We also have $\omega(f, \delta) \geq C_1 \delta \log (1/\delta)$.

(2) The Zygmund Class

It was Zygmund who brought (2) and (3) to a middle ground by finding a class of functions which includes lip 1 and has the property that membership in the class is equivalent to being able to be approximated (by $\mathcal{T}_N$) to within C/N.

Definition. $f \in Z$ iff $\|f(x + h) + f(x - h) - 2f(x)\| \leq Ch$ for any $h \geq 0$. We note that if $f \in \text{lip } 1$, then $f \in Z$. Thus, Z is an extension of the class lip 1.

THEOREM 4. $f \in Z$ iff $E_N^*(f) \leq C/N$.

Proof. Assume $f \in Z$. Let $K_n(\theta)$ be the Jackson kernel. That is

$$K_n(\theta) = \frac{1}{\lambda} \left(\frac{\sin \left(\frac{1}{2} \left[\frac{n}{2} \right] + \frac{1}{2} \right) \theta}{\sin \frac{\theta}{2}} \right)^4$$

where λ is such that $\int_{-\pi}^{\pi} K_n(\theta) d\theta = 1$. We have previously proved that

$\int_{-\pi}^{\pi} \theta K_n(\theta) d\theta \leq C_1/n$. From elementary knowledge of convolutions we know $f(x) - \int_{-\pi}^{\pi} f(x - \theta) K_n(\theta) d\theta = f(x) - \int_{-\pi}^{\pi} f(x + \theta) K_n(\theta) d\theta$. Therefore,

$$\left| f(x) - \int_{-\pi}^{\pi} f(x - \theta) K_n(\theta) d\theta \right|$$
$$= \frac{1}{2} \left| \int_{-\pi}^{\pi} (2f(x) - f(x - \theta) - f(x + \theta)) K_n(\theta) d\theta \right|$$
$$\leq \frac{C}{2} \int_{-\pi}^{\pi} |\theta| K_n(\theta) d\theta = C \int_0^{\pi} \theta K_n(\theta) d\theta \leq \frac{C_2}{n}$$

and thus $E_N^*(f) \leq C_2/N$.

Now assume $E_N^*(f) \leq C/N$. As before,

$$f(x) = \sum_{n=0}^{\infty} Q_n(x)$$

where Q_n is of degree 2^n and

$$\|Q_n\| \leq \frac{C}{2^n}.$$

We have

$$|Q_n(x + h) - 2Q_n(x) + Q_n(x - h)| = |h| |Q_n'(\xi_1) - Q_n'(\xi_2)|$$
$$\leq 2h^2 \|Q_n''\| \leq 2Ch^2 2^n$$

Let N be an arbitrary integer. Then

$$|f(x + h) - 2f(x) + f(x - h)|$$
$$= \left| \sum_{n=0}^{\infty} Q_n(x + h) - 2Q_n(x) + Q_n(x - h) \right|$$
$$\leq \sum_{n=0}^{N} |Q_n(x + h) - 2Q_n(x) + Q_n(x - h)|$$
$$+ 4 \sum_{n=N+1}^{\infty} \|Q_n\|$$
$$\leq 2h^2 C \sum_{n=0}^{N} 2^n + 4C \sum_{n=N+1}^{\infty} \left(\frac{1}{2^n} \right)$$
$$\leq 2h^2 C 2^{N+1} + \frac{4C}{2^N}.$$

Since we can assume $0 \leq h \leq 1$, we can pick N such that $1/h \leq 2^N \leq 2/h$. Then $2h^2 C 2^{N+1} + 4C/2^N \leq 8Ch + 4Ch = 12Ch$ and thus $f \in Z$. ∎

(3) Approximating Lip α

Let $0 < \alpha < 1$ and assume $f \in \text{lip } \alpha$. If $\alpha' < \alpha$ it is easy to see $f \in \text{lip } \alpha'$. From the fact that $f \in \text{lip } \alpha$ we have that $E_N^*(f) \leq C/N^\alpha$. Thus $N^{\alpha'} E_N^*(f) \leq N^{\alpha'} C/N^\alpha$ which approaches 0.

Thus, for any $\alpha' > 0$, we can find $f \in \text{lip } \alpha'$ such that $N^{\alpha'} E_N^*(f)$ approaches 0. We are thus led to the following question "Is C/N^α the correct way of approximating functions in lip α?" This question is answered by the following

THEOREM 5. *For all $\alpha \in (0, 1)$ there is an $f \in \text{lip } \alpha$ such that*

$$N^\alpha E_N^*(f) \geq C(\alpha) > 0.$$

Proof. Let $f(x) = \sum_{n=1}^{\infty} \dfrac{\cos(A^n x)}{A^{\alpha n}}$ where A is chosen so that $A^\alpha > 2$.

We first prove: $f \in \text{lip } \alpha$. For any δ we have

$$\cos[A^n(x + \delta)] - \cos A^n x = A^n \delta \sin \xi$$

for some ξ and thus $|\cos(A^n(x + \delta)) - \cos A^n x| \leq A^n \delta$. Let K be any positive integer. Then

$$|f(x + \delta) - f(x)| = \left| \sum_{n=1}^{\infty} \frac{\cos(A^n(x + \delta)) - \cos A^n x}{A^{\alpha n}} \right|$$

$$\leq \left| \sum_{n=1}^{K} \frac{\cos A^n(x + \delta) - \cos A^n x}{A^{\alpha n}} \right|$$

$$+ \left| \sum_{n=K+1}^{\infty} \frac{\cos A^n(x + \delta) - \cos A^n x}{A^{\alpha n}} \right|$$

$$\leq \sum_{n=1}^{K} \frac{A^n \delta}{A^{\alpha n}} + \sum_{n=K+1}^{\infty} \frac{2}{A^{\alpha n}}$$

$$\leq C_1 (\delta A^{K(1-\alpha)} + A^{-K\alpha}).$$

Choose K such that $1/\delta \leq A^K < A/\delta$. Then

$$|f(x + \delta) - f(x)| \leq C_1(\delta A^{K(1-\alpha)} + A^{-K\alpha}) \leq C_2 \delta^\alpha$$

and thus $f \in \text{lip } \alpha$.

We now prove: f can not be approximated better than C/N^α. Let $T_{A^n - 1}$ be a trigonometric polynomial $\in \mathscr{T}_{A^n - 1}$. Then

$$f(x) - T_{A^n - 1}(x) = Q(x) + \frac{\cos A^n x}{A^{\alpha n}} + \sum_{K=n+1}^{\infty} \frac{\cos A^K x}{A^{K\alpha}}$$

where $Q(x)$ is $\in \mathscr{T}_{A^n - 1}$. We observe that

$$\left| \sum_{K=n+1}^{\infty} \frac{\cos A^K x}{A^{K\alpha}} \right| \leq \sum_{K=n+1}^{\infty} \frac{1}{A^{K\alpha}} = \frac{1}{A^{n\alpha}(A^\alpha - 1)}$$

while

$$\left| \frac{\cos A^n x}{A^{\alpha n}} \right| = \frac{1}{A^{\alpha n}}$$

at $2A^n$ different places in $[0, 2\pi)$. At those places

$$\left| \frac{\cos A^n x}{A^{\alpha n}} + \sum_{K=n+1}^{\infty} \frac{\cos A^n x}{A^{\alpha n}} \right|$$

$$\geq \frac{1}{A^{\alpha n}} \left[1 - \frac{1}{A^\alpha - 1} \right] = \frac{1}{A^{\alpha n}} \left[\frac{A^\alpha - 2}{A^\alpha - 1} \right]$$

Since $Q(x)$ is $\in \mathcal{T}_{A^n-1}$, $Q(x)$ has at most $2(A^n - 1)$ zeroes in $[0, 2\pi)$. Therefore, at one of the places where $|\cos A^n x / A^{\alpha n}| = 1/A^{\alpha n}$, $Q(x)$ must have the same sign as $\cos A^n x / A^{\alpha n}$.

Therefore,

$$\|f - T_{A^n-1}\| \geq \frac{1}{A^{\alpha n}} \left[\frac{A^\alpha - 2}{A^\alpha - 1} \right]$$

Now, for any integer $m \geq 0$, let n be such that $A^{n-1} \leq m \leq A^n - 1$. Then

$$E_m^*(f) \geq E_{A^n-1}^*(f) = \inf_{T \in \mathcal{T}_{A^n-1}} \|f - T\| \geq \frac{1}{A^{\alpha n}} \left[\frac{A^\alpha - 2}{A^\alpha - 1} \right]$$

$$\geq \frac{1}{(Am)^\alpha} \left[\frac{A^\alpha - 2}{A^\alpha - 1} \right] = \frac{C(\alpha)}{m^\alpha}$$

Thus C/N^α is the best we can say about the order of approximation for the entire class lip α. ∎

CHAPTER VI

LINEAR OPERATORS

Theorem 3.9 proved that $E_n^*(\mathscr{S}^*) \geq \pi/2(n+1)$ (actually in Theorem 4.13 we established equality). That is, for each n, we found a function $f_n(x)$ such that $f_n(x) \in \mathscr{S}^*$ and $E_n^*(f_n) \geq \pi/2(n+1)$. However, can we find a function which will work for all n? To be more specific, does there exist a function $f \in \mathscr{S}^*$ and a $C > 0$ such that $E_n^*(f) \geq C/n$ for all n?

To find such a function we need to define a new method of summability for Fourier series.

(1) Summation of Fourier Series

We recall that for $f \in C^*[-\pi, \pi]$, if $f \sim \sum C_n e^{inx}$ then

$$S_N(f, x) = \sum_{n=-N}^{N} C_n e^{inx}$$

while

$$\sigma_N(f, x) = \sum_{n=-N}^{N} \left(1 - \frac{|n|}{N}\right) C_n e^{inx}$$

If we look at these two methods of summation pictorially, we find that

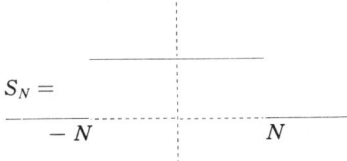

while

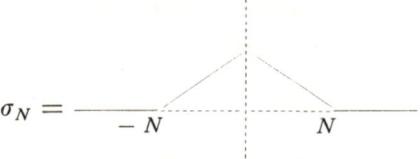

The new method of summability, D_N, can be described by

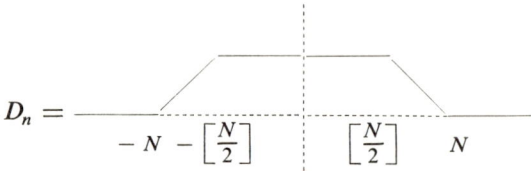

or

$$D_N(f, x) = \sum_{n=-[N/2]}^{[N/2]} C_n e^{inx} + 2 \sum_{[N/2]<|n|\leq N} \left(1 - \frac{|n|}{N}\right) C_n e^{inx}.$$

We observe that

$$D_{2N}(f, x) = 2\sigma_{2N}(f, x) - \sigma_N(f, x)$$

and $D_{2N}(T_N, x) \equiv T_N(x)$ for $T_N(x) \in \mathcal{T}_N$. The next theorem proves that as N approaches infinity, $D_{2N}(f, x)$ converges uniformly to $f(x)$ for $f \in C^*[-\pi, \pi]$. We leave as an exercise the theorem that $D_{2N+1}(f, x)$ converges uniformly to $f(x)$ as N approaches infinity.

THEOREM 1. For any $x \in [-\pi, \pi]$ and any N,

$$|D_{2N}(f, x) - f(x)| \leq 4E_N^*(f).$$

Proof. Since

$$D_{2N}(f, x) = 2\sigma_{2N}(f, x) - \sigma_N(f, x)$$

and since $\sigma_N(f, x) = (f*K_N)(x)$ where $K_N(x)$, the Fejér kernel, has the properties that $K_N(x) \geq 0$ and $\int_{-\pi}^{\pi} K_N(\theta) d\theta = 1$, we have

$$D_{2N}(f, x) = 2\int_{-\pi}^{\pi} f(\theta) K_{2N}(x - \theta) d\theta - \int_{-\pi}^{\pi} f(\theta) K_N(x - \theta) d\theta.$$

Thus $|D_{2N}(f, x)| \leq 3\|f\|$. Now let $T_N(x) \in \mathcal{T}_N(x)$ be such that $\|f - T_N\| = E_N^*(f)$. Since $D_{2N}(T_N, x) \equiv T_N(x)$ we have

$$\begin{aligned}
|f(x) - D_{2N}(f, x)| &= |f(x) - T_N(x) - D_{2N}(f, x) + D_{2N}(T_N, x)| \\
&= |f(x) - T_N(x)) - D_{2N}(f - T_N, x)| \\
&\leq |f(x) - T_N(x)| + |D_{2N}(f - T_N, x)| \\
&\leq 4\|f - T_N\| = 4 E_N^*(f). \blacksquare
\end{aligned}$$

It is this theorem which we shall use to prove that $E_N^*(|\sin x|) \geq C/N$. We first compute the Fourier series of $|\sin x|$ and get

$$|\sin x| = \frac{2}{\pi} - \frac{4}{\pi}\left[\frac{\cos 2x}{2^2 - 1} + \frac{\cos 4x}{4^2 - 1} + \cdots\right]$$

with equality because of the uniform convergence of the Fourier series. We know that

$$4E_N^*(|\sin x|) \geq \|D_{2N}(|\sin x|, x) - |\sin x|\| \geq |D_{2N}(|\sin x|, 0) - |\sin 0||$$

$$= \left|\frac{4}{\pi}\sum_{\substack{n=N+1\\n\text{ even}}}^{2N}\left(\frac{n}{N} - 1\right)\left(\frac{1}{n^2 - 1}\right) + \frac{4}{\pi}\sum_{\substack{n=2N+1\\n\text{ even}}}^{\infty}\frac{1}{n^2 - 1}\right|$$

$$\geq \frac{4}{\pi}\sum_{j=N}^{\infty}\frac{1}{(2j)^2 - 1}$$

$$= \frac{2}{\pi(2N - 1)}$$

Thus $E_N^*(|\sin x|) \geq 1/2\pi(2N - 1) > 1/4\pi N$.

Note.

$$\||\cos x| - T_N(x)\| = \left\|\left|\sin\left(x - \frac{\pi}{2}\right)\right| - T_N(x)\right\|$$

$$= \left\|\left|(\sin \theta)\right| - T_N\left(\theta + \frac{\pi}{2}\right)\right\| > \frac{1}{4\pi N}$$

and thus $E_N^*(|\cos x|) > 1/4\pi N$.

We leave as an exercise the proof that

$$E_N(|x|) > \frac{1}{4\pi N}$$

(where $x \in [-1, 1]$).

The function $|\sin x|$ also provides us with another example. Namely, it is an example of a function for which the Cesàro sums do not converge to the function as quickly as we can approximate it by trigonometric polynomials. Since $|\sin x| \in \mathscr{S}^*$, we know $E_N^*(|\sin x|) \leq C/N$ for some C. However, with regards to Cesàro sums we have the following

THEOREM 2.

$$\|\sigma_N(|\sin x|, t) - |\sin t|\| \geq \frac{2 \log N}{\pi N}$$

Proof. It obviously would suffice to prove the statement only for $t = 0$. That is,

$$|\sigma_N(|\sin x|, 0) - |\sin 0|| \geq \frac{2 \log N}{\pi N}$$

or that

$$\sigma_N(|\sin x|, 0) \geq \frac{2 \log N}{\pi N}$$

For N even,

$$\begin{aligned}
\sigma_N(|\sin x|, 0) &= \frac{1}{N}\left[\frac{2N}{\pi} - \frac{4}{\pi}\left(\frac{N-2}{2^2-1} + \frac{N-4}{4^2-1} + \cdots + \frac{2}{(N-2)^2-1}\right)\right] \\
&= \frac{2}{\pi} - \frac{4}{\pi}\left(\frac{1}{2^2-1} + \frac{1}{4^2-1} + \cdots \frac{1}{(N-2)^2-1}\right) \\
&\quad + \frac{4}{\pi N}\left(\frac{2}{2^2-1} + \frac{4}{4^2-1} + \cdots + \frac{(N-2)}{(N-2)^2-1}\right) \\
&= \frac{2}{\pi} - \frac{2}{\pi}\left(1 - \frac{1}{N-1}\right) \\
&\quad + \frac{2}{\pi N}\left(1 + \frac{2}{3} + \frac{2}{5} + \cdots + \frac{2}{N-3} + \frac{1}{N-1}\right) \\
&= \frac{4}{\pi N}\left(1 + \frac{1}{3} + \frac{1}{5} + \cdots + \frac{1}{N-1}\right) \\
&\geq \frac{2 \log N}{\pi N}
\end{aligned}$$

For N odd, we obtain similar estimates for $\sigma_N(|\sin x|, 0)$. ∎

(2) Bounded Linear Operators

Returning to the bounded linear operators $D_N: C^*[-\pi, \pi] \to \mathcal{T}_N$ we note that they have the following two properties:

(1) $D_{2N}(T_N, x) \equiv T_N(X)$ for $T_N \in \mathcal{T}_N$.
(2) $\|D_N(f, x) - f(x)\|$ converges to zero as N approaches infinity, for all $f \in C^*[-\pi, \pi]$.

We would now like to see whether we can find bounded linear operators $L_N : C^*[-\pi, \pi] \to \mathcal{T}_N$ with the properties

(1') $L_N(T_N, X) \equiv T_N(X)$ for $T_N \in \mathcal{T}_N$
(2) $\|L_N(f, x) - f(x)\|$ converges to zero as N approaches infinity for all $f \in C^*[-\pi, \pi]$.

Before discussing the existence of a sequence of bounded linear operators L_N with properties (1′) and (2), we look at a specific example of bounded linear operators, the partial sums operators S_N. To each $f \in C^*[-\pi, \pi]$, S_N assigns the N-th partial sum of its Fourier series. Clearly S_N has property (1′).

We shall now prove that S_N does not have property (2). To prove this, we need the following lemma. In the lemma, by $\|S_N\|$ we mean the operator norm; i.e.

$$\|S_N\| = \sup_{\substack{f \in C^*[-\pi,\pi] \\ \|f\|=1}} \|S_N(f, x)\| = \sup_{\substack{f \in C^*[-\pi,\pi] \\ \|f\| \neq 0}} \frac{\|S_N(f, x)\|}{\|f\|}.$$

LEMMA 3.

$$\|S_n\| = \frac{1}{2\pi} \int_{-\pi}^{\pi} \left| \frac{\sin\left(n + \frac{1}{2}\right)t}{\sin \frac{1}{2}t} \right| dt = \frac{4}{\pi^2} \log n + 0(1).$$

Proof. Since

$$S_n(f, x) = \frac{1}{2\pi} \int_{-\pi}^{\pi} f(x - t) \frac{\sin\left(n + \frac{1}{2}\right)t}{\sin \frac{1}{2}t} dt$$

it is obvious that

$$|S_n(f, x)| \leq \frac{\|f\|}{2\pi} \int_{-\pi}^{\pi} \left| \frac{\sin\left(n + \frac{1}{2}\right)t}{\sin \frac{t}{2}} \right| dt$$

and consequently

$$\|S_n\| \leq \frac{1}{2\pi} \int_{-\pi}^{\pi} \left| \frac{\sin\left(n + \frac{1}{2}\right)t}{\sin \frac{1}{2}t} \right| dt$$

To prove the inequality the other way, we take a step function $g(x)$ which is 1 where

$$\frac{\sin\left(n + \frac{1}{2}\right)t}{\sin \frac{1}{2}t}$$

is non-negative and -1 where

$$\frac{\sin\left(n+\frac{1}{2}\right)t}{\sin\frac{1}{2}t}$$

is negative. For any $\delta > 0$ we can find an $f \in C^*[-\pi, \pi]$ with $\|f\| = 1$ such that $f(x) \equiv g(x)$ except on a set E of measure $< \delta$. Then,

$$|S_n(f, x)| = |S_n(f - g, x) + S_n(g, x)|$$
$$\geq |S_n(g, x)| - |S_n(f - g, x)|$$
$$= \frac{1}{2\pi} \int_{-\pi}^{\pi} \left|\frac{\sin\left(n+\frac{1}{2}\right)t}{\sin\frac{1}{2}t}\right| dt$$

$$- \frac{1}{2\pi} \int_E (f(x-t) - g(x-t)) \frac{\sin\left(n+\frac{1}{2}\right)t}{\sin\frac{1}{2}t} dt$$

$$\geq \frac{1}{2\pi} \int_{-\pi}^{\pi} \left|\frac{\sin\left(n+\frac{1}{2}\right)t}{\sin\frac{1}{2}t}\right| dt$$

$$- \frac{1}{\pi} \int_E \left|\frac{\sin\left(n+\frac{1}{2}\right)t}{\sin\frac{1}{2}t}\right| dt$$

$$\geq \frac{1}{2\pi} \int_{-\pi}^{\pi} \left|\frac{\sin\left(n+\frac{1}{2}\right)t}{\sin\frac{1}{2}t}\right| dt - \frac{1}{\pi}(2n+1)\delta$$

By taking δ small we can come as close as we want to

$$\frac{1}{2\pi} \int_{-\pi}^{\pi} \left|\frac{\sin\left(n+\frac{1}{2}\right)t}{\sin\frac{1}{2}t}\right| dt.$$

Thus

$$\sup_{\substack{f \in C^*[-\pi,\pi] \\ \|f\|=1}} \|S_n(f,x)\| \geq \frac{1}{2\pi} \int_{-\pi}^{\pi} \left| \frac{\sin\left(n+\frac{1}{2}\right)t}{\sin\frac{1}{2}t} \right| dt$$

and we get equality.

To estimate

$$\frac{1}{2\pi} \int_{-\pi}^{\pi} \left| \frac{\sin\left(n+\frac{1}{2}\right)t}{\sin\frac{1}{2}t} \right| dt$$

we proceed as follows. We first change variables and get

$$\frac{2}{\pi} \int_0^{\pi/2} \left| \frac{\sin(2n+1)t}{\sin t} \right| dt.$$

If we forget about $\int_0^{\pi/(2n+1)}$ and $\int_{n\pi/(2n+1)}^{\pi/2}$ we have only incurred a bounded error. If we then replace $1/\sin t$ by $1/t$ we again incur only a bounded error since

$$\left| \frac{1}{\sin t} - \frac{1}{t} \right|$$

is bounded. Hence we have

$$\frac{2}{\pi} \int_{\pi/(2n+1)}^{n\pi/(2n+1)} \left| \frac{\sin(2n+1)t}{t} \right| dt + 0(1)$$

We break the integral up into

$$\int_{\pi/(2n+1)}^{2\pi/(2n+1)} + \int_{2\pi/(2n+1)}^{3\pi/(2n+1)} + \ldots + \int_{(n-1)\pi/(2n+1)}^{n\pi/(2n+1)}$$

so that $\sin(2n+1)t$ is of constant sign in each interval. In estimating these integrals from above, we replace $1/t$ by the reciprocal of the left end point and in estimating them from below, we use the right end point. Since

$$\int_{K\pi/(2n+1)}^{(K+1)\pi/(2n+1)} |\sin(2n+1)t| dt \doteq \frac{2}{2n+1}$$

we have

$$\frac{2}{\pi} \int_{\pi/(2n+1)}^{n\pi/(2n+1)} \left| \frac{\sin(2n+1)t}{\sin t} \right| dt$$

estimated from above by

$$\frac{4}{\pi^2}\left[1 + \frac{1}{2} + \cdots + \frac{1}{n-1}\right]$$

and from below by

$$\frac{4}{\pi^2}\left[\frac{1}{2} + \frac{1}{3} + \cdots + \frac{1}{n}\right]$$

both of which are $4/\pi^2 \log n + 0(1)$. ∎

To prove $\{S_N\}$ does not have property (2) we use the famous

Uniform Boundedness Theorem: Let $\{L_n\}$ be a sequence of bounded linear operators mapping a Banach space X into a Banach space Y. Then for each $x \in X$, $\{\|L_n(x)\|\}$ is bounded iff $\{\|L_n\|\}$ is bounded.

To apply this theorem to our case, we let $\{S_N\}$ be the sequence of bounded linear operators, $X = C^*[-\pi, \pi]$ and $Y = \mathcal{T}_N$. Since $\{\|S_N\|\}$ is unbounded, there is an $f \in C^*[-\pi, \pi]$ such that $\{\|S_N(f)\|\}$ is unbounded. Obviously, for that f, $\|S_N(f, x) - f(x)\|$ is unbounded. Hence, we have the following

THEOREM 4. *There is a function $f \in C^*[-\pi, \pi]$ whose Fourier series does not converge uniformly.*

We can apply the Uniform Boundedness Theorem in a different way and obtain an even stronger theorem. We pick any $x_0 \in [-\pi, \pi]$ and let S_N be the operator taking $f \in C^*[-\pi, \pi]$ to the N-th partial Fourier sum of $f(x)$ evaluated at x_0; i.e., $S_N: f \to S_N(f, x_0)$ (even though S_N is a slightly different operator, $\|S_n\|$ is the same).

Since $\|S_N\|$ is unbounded, there is an $f \in C^*[-\pi, \pi]$ such that $|S_N(f, x_0)|$ is unbounded. We therefore have the following

THEOREM 5. *Given any $x_0 \in [-\pi, \pi]$, there is an $f \in C^*[-\pi, \pi]$ such that the Fourier sums of f at x_0 are unbounded.*

One criterion for the convergence of the Fourier series is the well known Dini-Lipschitz theorem: If $f \in C^*[-\pi, \pi]$ and $\omega(f, |\delta|) \log |\delta|$ goes to 0 as δ does, the Fourier series of f converges uniformly to $f(x)$.

Proof. Let $T_N \in \mathcal{T}_N$ be such that $\|f - T_N\| = E_N^*(f) \leq C\omega(f, 1/N)$. Then,

$$\|S_N(f, x) - f(x)\| = \|S_N(f - T_N, x) - (f(x) - T_N(x))\|$$
$$\leq \|f - T_N\|[\|S_N\| + 1]$$
$$\leq C\omega\left(f, \frac{1}{N}\right)\left[\frac{4}{\pi^2} \log N + 0(1)\right]$$

which goes to zero as N approaches infinity. ∎

Corollary. If $f \in \text{lip } \alpha$, then the Fourier series of $f(x)$ converges uniformly to $f(x)$.

We will now prove the non-existence of a linear operator with properties (1′) and (2). For this we need the following

LEMMA 6. *Let L_N be a linear operator from $C^*[-\pi, \pi]$ into $\mathcal{T}_N$ such that $L_N(T_N) \equiv T_N$ for $T_N \in \mathcal{T}_N$. Then $\|L_N\| \geq \|S_N\|$.*

Proof. Define the following linear operators: For any real α, let $\cup^{\alpha}(f(x)) = f(x+\alpha)$. Thus $\cup^{\alpha}: C^*[-\pi, \pi] \to C^*[-\pi, \pi]$. It is obvious that $\|\cup^{\alpha}\| = 1$.

Claim.

i.e.
$$S_N = \frac{1}{2\pi} \int_{-\pi}^{\pi} \cup^{-\alpha} L_N \cup^{\alpha} d\alpha$$

$$S_N(f, x) = \frac{1}{2\pi} \int_{-\pi}^{\pi} \cup^{-\alpha}(L_N(\cup^{\alpha}(f(x)))) \, d\alpha .$$

Once we prove the claim, the lemma is easily proven as

$$|S_N(f, x)| \leq \frac{1}{2\pi} \int_{-\pi}^{\pi} |\cup^{-\alpha}(L_N(\cup^{\alpha}(f(x))))| \, d\alpha$$

$$\leq \frac{1}{2\pi} \int_{-\pi}^{\pi} \|\cup^{-\alpha}\| \|L_N\| \|\cup^{\alpha}\| \|f\| \, d\alpha$$

$$= \|L_N\| \|f\|$$

and thus $\|S_N\| \leq \|L_N\|$. To prove the claim, set

$$L(f(x)) = \frac{1}{2\pi} \int_{-\pi}^{\pi} \cup^{-\alpha}(L_N(\cup^{\alpha}(f(x)))) \, d\alpha$$

Let $|m| \leq N$.

$$L(e^{imx}) = \frac{1}{2\pi} \int_{-\pi}^{\pi} \cup^{-\alpha}(L_N(e^{im(x+\alpha)}) \, d\alpha$$

$$= \frac{1}{2\pi} \int_{-\pi}^{\pi} \cup^{-\alpha}(e^{im(x+\alpha)}) \, d\alpha$$

$$= \frac{1}{2\pi} \int_{-\pi}^{\pi} e^{imx} d\alpha$$

$$= e^{imx} .$$

Now let $|m| > N$.

$$L(e^{imx}) = \frac{1}{2\pi} \int_{-\pi}^{\pi} \cup^{-\alpha}(L_N(\cup^{\alpha}(e^{imx}))) \, d\alpha$$

$$= \frac{1}{2\pi} \int_{-\pi}^{\pi} \cup^{-\alpha}(L_N(e^{im(x+\alpha)})) \, d\alpha$$

$$= \frac{1}{2\pi} \int_{-\pi}^{\pi} \cup^{-\alpha}(\sum_{k=-N}^{N} C_k e^{ikx}) e^{im\alpha} \, d\alpha$$

$$= 0$$

Therefore $L(e^{imx}) = S_N(e^{imx})$ for any m. Now let $f \in C^*[-\pi, \pi]$ and find $T_K \in \mathcal{T}_K$ such that $\|f - T_K\| < \varepsilon$ where $\varepsilon > 0$ is given. Then

$$|L(f, x) - S_N(f, x)| = |L(f - T_K, x) - S_N(f - T_K, x)|$$
$$\leq \|f - T_K\|[\|L\| + \|S_N\|]$$
$$\leq \varepsilon\|L_N\| + \varepsilon\|S_N\| \quad (\text{Since } \|L\| \leq \|L_N\|).$$

Therefore $L(f, x) \equiv S_N(f, x)$ and the claim (and thereby the lemma) is proven. ∎

We can now prove the aforementioned

THEOREM 7. There does not exist a sequence of bounded linear transformations $\{L_N\}$ such that

(1') $L_N(T_N(x)) \equiv T_N(x)$ for $T_N \in \mathcal{T}_N$.

(2) $\|L_N(f) - f\|$ converges to 0 for all $f \in C^*[-\pi, \pi]$.

Proof. By the previous lemma, since $\{L_N\}$ has property (1')

$$\|L_N\| \geq \|S_N\| = 4/\pi^2 \log N + 0(1).$$

By the Uniform Boundedness Theorem, there is an $f \in C^*[-\pi, \pi]$ such that $\{\|L_N(f)\|\}$ is unbounded. Hence $\{L_n\}$ does not have property (2). ∎

Consider the operator T_N which assigns to each $f \in C^*[-\pi, \pi]$ its best N-th degree trigonometric approximant. We know $\{T_N\}$ has properties (1') and (2). Since

$$\|T_N(f)\| \leq \|T_N(f) - f\| + \|f\|$$
$$\leq C\omega\left(f, \frac{1}{N}\right) + \|f\|$$
$$\leq 2C\|f\| + \|f\|$$
$$= \|f\|[2C + 1]$$

we know that $\{T_N\}$ are bounded operators. Hence, we must conclude that T_N is not linear. This can also be seen from the following example. Let $f(x)$ be defined as

$$f(x) = \begin{cases} 0 & x = -1 \\ 1 & x = -\frac{1}{2} \\ 0 & x = 0 \\ 1 & x = \frac{K}{2N} \quad K \text{ odd}, \ 1 \leq K < 2N \\ -1 & x = \frac{2N}{K} \quad K \text{ even}, \ 2 \leq K < 2N \\ 0 & x = 1 \\ \text{linear between these points} \end{cases}$$

Then obviously $E_N^*(f) = 1$ and $T_N(f) \equiv 0$. Define

$$g(x) = \begin{cases} f(x) & x < 0 \\ -f(x) & x \geq 0 \end{cases}$$

Then $E_N^*(g) = 1$ and $T_N(g) \equiv 0$.

Then $f(x) + g(x)$ looks like

If T_N were linear, then $T_N(f + g) \equiv 0$ and we would have

$$E_N^*(f + g) = 2.$$

However, $|f(x) + g(x) - 1| \leq 1$, and thus $E_N^*(f + g) \leq 1$. Consequently T_N is not linear.

This same example can be used to prove that the operator assigning to each $f \in C[-1, 1]$ its best N-th degree polynomial approximant is also non-linear.

We leave as an exercise the proof of the following

THEOREM 8. Let $\{L_N\}$ be a sequence of bounded linear operators such that L_N maps $C[a, b]$ into P_N. Then, if $L_N(P_N(x)) \equiv P_N(x)$ for all $P_N \in P_N$ and all N, there is a $f \in C[a, b]$ such that $\|L_N f - f\|$ does not go to zero.

(3) Korovkin's Theorem

We have just seen that for bounded linear operators $\{L_N\}$ such that $L_N: C[a, b] \to \mathscr{P}_N$ $L_N(P_N) = P_N$ implies the existence of an f such that $\|L_N f - f\|$ does not go to zero. Korovkin changed the hypothesis and obtained a necessary and sufficient condition for $\|L_N f - f\|$ to go to 0 for all $f \in C[a, b]$. The change was that he specified that the operators $\{L_N\}$ were positive; i.e., if $f(t) \geq 0$ for all $t \in [a, b]$, then $L_N(f, t) \geq 0$ for all $t \in [a, b]$.

KOROVKIN'S THEOREM. Let $\{L_N\}$ be a sequence of positive linear bounded operators such that $L_N: C[a, b] \to P_N$. Then $\|L_N f - f\|$ goes to 0 for all $f \in C$ iff $\|L_N x^i - x^i\|$ goes to 0 for $i = 0, 1, 2$.

There is an analogue of this theorem for periodic functions with $\mathscr{T}_N$ instead of $\mathscr{P}_N$, $C^*[-\pi, \pi]$ instead of $C[a, b]$ and $\{1, \cos x, \sin x\}$ instead of $\{1, x, x^2\}$.

A recent paper by Shisha and Mond (see [31]) asked (and answered) the following natural question "How quickly does $\|L_N(f) - f\|$ go to zero?" We will state and prove the answer to this question (and thereby prove Korovkin's theorem).

Notation. By $L_N((t - x)^2, x)$ we mean that x is fixed in $[a, b]$ while t forms the functions $1, t, t^2$ on which L_N operates. That is,

$$L_N((t - x)^2, x) = L_N(t^2 - 2tx + x^2, x)$$
$$= L_N(t^2, x) - 2xL_N(t, x) + x^2 L_N(1, x).$$

THEOREM 9. Let $\{L_N\}$ be a sequence of positive, linear bounded operators taking $C[a, b] \to \mathscr{P}_N$. Then, if $f \in C[a, b]$ (and $\omega(f, \cdot)$ is its modulus of continuity),

$$\|L_N(f) - f\| \leq \|f\| \cdot \|L_N(1) - 1\| + \|L_N(1) + 1\| \omega(f, u_N)$$

where $u_N = \|L_N((t - x)^2, x\|^{1/2}$. If $L_N(1) = 1$, as is frequently the case, this reduces to $\|L_N(f) - f\| \leq 2\omega(f, u_N)$.

Note. If $\|L_N x^i - x^i\|$ goes to zero for $i = 0, 1, 2$, then u_N goes to zero and since $\omega(f, 0^+) = 0$, we get that $\|L_N f - f\|$ goes to zero for all $f \in C[a, b]$. Thus theorem 9 contains Korovkin's theorem.

Proof of theorem 9. Let $x_0 \in [a, b]$, let δ be positive and let x always be in $[a, b]$. If $|x - x_0| > \delta$, then,

$$|f(x) - f(x_0)| \leq \omega(f, |x - x_0|) = \omega\left(f, \frac{|x - x_0|\delta}{\delta}\right)$$

$$\leq \left(1 + \frac{|x - x_0|}{\delta}\right)\omega(f, \delta)$$

$$\leq \left(1 + \frac{(x - x_0)^2}{\delta^2}\right)\omega(f, \delta)$$

The inequality,

$$|f(x) - f(x_0)| \leq \left(1 + \frac{(x - x_0)^2}{\delta^2}\right)\omega(f, \delta)$$

obviously holds for $|x - x_0| \leq \delta$ also.

Since $L_N(f(x_0), x) = f(x_0)L_N(1, x)$ we have

$$|L_N(f, x) - f(x_0)L_N(1, x)|$$
$$= |L_N(f(x) - f(x_0), x)|$$
$$\leq \left|L_N\left(\left(1 + \frac{(x - x_0)^2}{\delta^2}\right)\omega(f, \delta), x\right)\right|$$
$$= \omega(f, \delta)\left[L_N(1, x) + \frac{1}{\delta^2}L_N((x - x_0)^2, x)\right]$$
$$\leq \omega(f, \delta)\left[L_N(1, x) + \left(\frac{u_N}{\delta}\right)^2\right]$$

Setting $x = x_0$, we get

$$|L_N(f, x_0) - f(x_0)L_N(1, x_0)| \leq \omega(f, \delta)\left[L_N(1, x_0) + \left(\frac{u_N}{\delta}\right)^2\right]$$

We also have,

$$|-f(x_0) + f(x_0)L_N(1, x_0)| \leq \|f\|\|-1 + L_N(1, x)\|.$$

Combining these two inequalities, we get

$$|L_N(f, x_0) - f(x_0)|$$
$$\leq \|f\|\|L_N(1, x) - 1\| + \omega(f, \delta)\left[L_N(1, x_0) + \left(\frac{u_N}{\delta}\right)^2\right]$$

If $u_N > 0$, we set $\delta = u_N$ and get the inequality claimed in theorem 9. If $u_N = 0$, our inequality becomes

$$|L_N(f, x_0) - f(x_0)| \leq \|f\|\|L_N(1, x) - 1\| + \omega(f, \delta)L_N(1, x_0).$$

We then let δ approach zero, and since $\omega(f, 0^+) = 0$, we get

$$|L_N(f, x_0) - f(x_0)| \leq \|f\|\|L_N(1, x) - 1\|$$

which is what the inequality of theorem 9 reduces to for $u_N = 0$. ∎

Example. Let $[a,b] = [0,1]$ and let

$$L_N(f,x) = \sum_{k=0}^{N} \binom{N}{k} f\left(\frac{k}{N}\right) x^k (1-x)^{N-k}$$

That is, L_N is the Bernstein operator.

$$L_N(1,x) \equiv 1$$
$$L_N(t,x) \equiv x$$
$$L_N(t^2,x) = \frac{(N-1)x^2}{N^2} + \frac{x}{N}.$$

Thus

$$L_N((t-x)^2,x) = \frac{x-x^2}{N} \quad \text{and} \quad u_N = \frac{1}{\sqrt{4N}}.$$

By theorem 9, we get

$$|L_N(f,x) - f(x)| \leq 2\omega\left(f, \frac{1}{2\sqrt{N}}\right) \leq 2\omega\left(f, \frac{1}{\sqrt{N}}\right).$$

We thus have the well known result that the Bernstein polynomials converge to the function at the rate of $1/\sqrt{N}$.

For trigonometric polynomials we have a similar theorem.

THEOREM 10. *Let $\{L_N\}$ be a sequence of bounded, linear positive operators taking $C^*[-\pi,\pi] \to \mathcal{T}_N$. Then for $f \in C^*[-\pi,\pi]$,*

$$|L_N(f,x) - f(x)| \leq \|f\| \|L_N(1,x) - 1\| + \|L_N(1,x) + 1\| \omega(f,u_N)$$

where

$$u_N = \pi \left\| L_N\left(\sin^2\left(\frac{t-x}{2}\right), x\right) \right\|^{1/2}$$

If $L_N(1,x) \equiv 1$, then we have $|L_N(f,x) - f(x)| \leq 2\omega(f,u_N)$.

Example. Let $L_N(f,x) = (f*K_N)(x)$ where K_N is the Jackson kernel of degree N. One can easily show that $L_N(1,x) \equiv 1$ and

$$u_N = \pi \left[\frac{3}{4N^2+2}\right]^{1/2} \leq \frac{3}{N}.$$

Thus $|f(x) - L_N(f,x)| \leq 2\omega(f,u_N) \leq 6\omega(f,(1/N))$ and we have another proof of the Jackson theorem for trigonometric polynomials.

CHAPTER VII

RATIONAL APPROXIMATION

We have proven various theorems about how well we can approximate functions by polynomials. We now extend our class of approximating functions to rational polynomials and we ask the same kinds of questions. As we shall see, while for some functions we can get better approximations than with polynomials, we can't do any better for some classes of functions.

Definition. By a rational polynomial of degree n or less, we mean something of the form $P(x)/Q(x)$ where both $P(x)$ and $Q(x)$ are real polynomials of degree less than or equal to n. $\mathscr{R}_n$ will denote the entire class of rational polynomials of degree n or less.

Definition. By $R_n(f)$ we mean $\inf_{R \in \mathscr{R}_n} \|f - R\|$.

Definition. For C, a class of functions, by $R_n(C)$ we mean $\sup_{f \in C} R_n(f)$.

(1) Rational vs. Polynomials

Since $\mathscr{P}_n \subset \mathscr{R}_n$ it is obvious that $R_n(f) \leq E_n(f)$ and $R_n(C) \leq E_n(C)$. It is also obvious that for $f(x)$ such that $f \in \mathscr{R}_N$ but $f \notin \mathscr{P}_N$, that $R_n(f) = 0$ for $n \geq N$ while $E_n(f) > 0$ and thus $R_n(f) < E_n(f)$.

However, as we stated, for many classes C of functions, $R_n(C)$ behaves like $E_n(C)$. This is evident in the following theorems.

THEOREM 1. $R_n(\mathscr{S}) \geq 1/2(n+1)$.

Proof. We take the same function that was used to prove

$$E_n(\mathscr{S}) \geq \frac{1}{2(n+1)}.$$

That is:

$$f\left(\frac{K}{n+1}\right) = \frac{(-1)^K}{2(n+1)}, \quad K = 0, \ldots, n+1$$

and $f(x)$ is linear in between. Then $f(x)$ alternates sign at the $n+2$ points $\{K/(n+1)\}_{K=0}^{n+1}$ and since $P(x)/Q(x)$ only alternates signs when $P(x)$ does (Q can never be 0 in $[0, 1]$), $P(x)/Q(x)$ can only alternate signs at most $n+1$ times. Thus, for any $R \in \mathscr{R}_n$ at at least one of the points of $\{K/(n+1)\}_{K=0}^{n+1}$, $R(x)$ and $f(x)$ will have different signs and thus

$$\|f(x) - R(x)\| \geq \frac{1}{2(n+1)}.$$

Consequently

$$R_n(f) = \inf_{R \in \mathscr{R}_n} \|f - R\| \geq \frac{1}{2(n+1)}. \quad \blacksquare$$

Of course, since $\mathscr{P}_n \subset \mathscr{R}_n$, we must have $R_n(\mathscr{S}) \leq C/n$ by Jackson's theorem. Thus the correct order of magnitude for $R_n(\mathscr{S})$ is $1/n$, the same as it is for $E_n(\mathscr{S})$.

We shall now prove that the same thing happens for the class lip α ($0 < \alpha < 1$). We shall prove the existence of an $f(x) \in$ lip α such that $R_n(f) \geq C_1/n^\alpha$ and since, by Jackson's theorem we know $R_n(f) \leq E_n(f) \leq C_2/n^\alpha$ we will get that $1/n^\alpha$ is the correct order of magnitude for both $E_n(f)$ and $R_n(f)$ for that $f \in$ lip α. (The counterexample is modeled after that of theorem 5.5.)

THEOREM 2. Let $0 < \alpha < 1$. Then there exists an $f(x) \in$ lip α such that $n^\alpha R_n(f) \geq C(\alpha) > 0$.

Proof. Let A be a multiple of 8 such that $A^{2\alpha} > 2$. Let

$$f(x) = \sum_{n=1}^{\infty} \frac{\cos\left(A^n \arccos\left(\frac{x}{\sqrt{2}}\right)\right)}{A^{\alpha n}} = \sum_{n=1}^{\infty} \frac{T_{A^n}\left(\frac{x}{\sqrt{2}}\right)}{A^{\alpha n}}.$$

We first prove: (a) $f \in$ lip α. We notice that $\|T_{A^n}(x/\sqrt{2})\| \leq 1$ and

$$\left|\left[\cos\left(A^n \arccos\left(\frac{x}{\sqrt{2}}\right)\right)\right]'\right| = \left|\frac{\sin\left(A^n \arccos\left(\frac{x}{\sqrt{2}}\right)\right) A^n}{\sqrt{2}\sqrt{1 - \left(\frac{x}{\sqrt{2}}\right)^2}}\right| \leq A^n$$

for $x \in [0, 1]$. Let K be any positive integer, let $x \in [0, 1]$ and $\delta \in (-1, 1)$. Then

RATIONAL APPROXIMATION

$$|f(x+\delta)-f(x)| \leq \sum_{n=1}^{\infty} \left| \frac{T_{A^n}\left(\frac{x+\delta}{\sqrt{2}}\right) - T_{A^n}\left(\frac{x}{\sqrt{2}}\right)}{A^{\alpha n}} \right|$$

$$= \sum_{n=1}^{K} \left| \frac{T_{A^n}\left(\frac{x+\delta}{\sqrt{2}}\right) - T_{A^n}\left(\frac{x}{\sqrt{2}}\right)}{A^{\alpha n}} \right|$$

$$+ \sum_{n=K+1}^{\infty} \left| \frac{T_{A^n}\left(\frac{x+\delta}{\sqrt{2}}\right) - T_{A^n}\left(\frac{x}{\sqrt{2}}\right)}{A^{\alpha n}} \right|$$

$$\leq \sum_{n=1}^{K} \left(\frac{|\delta|}{\sqrt{2}}\right) \left| \frac{\sin\left(A^n \arccos\left(\frac{\xi_n}{\sqrt{2}}\right)\right) \cdot A^n}{\sqrt{2}\sqrt{1-\left(\frac{\xi_n}{\sqrt{2}}\right)^2} A^{\alpha n}} \right|$$

$$+ \sum_{n=K+1}^{\infty} \frac{2}{A^{\alpha n}}$$

$$\leq \sum_{n=1}^{K} \frac{|\delta| A^{n(1-\alpha)}}{\sqrt{2}} + \frac{2}{A^{\alpha n}} \frac{1}{A^\alpha - 1}$$

$$\leq C_1(|\delta| A^{K(1-\alpha)} + A^{-\alpha K}).$$

Now choose K such that A^K is about $1/|\delta|$ or $K = [-\log|\delta|/\log A]$. Then

$$A^{K(1-\alpha)} \leq |\delta|^{\alpha-1} \quad \text{and} \quad A^{-\alpha K} \leq A^\alpha |\delta|^\alpha.$$

Thus

$$|f(x+\delta) - f(x)| \leq C_1(|\delta|^\alpha + A^\alpha |\delta|^\alpha) = C_2 |\delta|^\alpha$$

and

$$f \in \mathrm{lip}\,\alpha.$$

We now prove: (b) $n^\alpha R^n(f) \geq C(\alpha) > 0$.
Consider any $R(x) \in \mathscr{R}_{A^N-1}$. Then

$$\|f(x) - R(x)\| \geq \left\| \sum_{n=1}^{N-1} \frac{T_{A^n}\left(\frac{x}{\sqrt{2}}\right)}{A^{\alpha n}} - R(x) + \frac{T_{A^N}\left(\frac{x}{\sqrt{2}}\right)}{A^{\alpha N}} \right\|$$

$$- \sum_{n=N+1}^{\infty} \frac{1}{A^{\alpha n}}$$

$$= \left\| \sum_{n=1}^{N-1} \frac{T_{A^n}\left(\frac{x}{\sqrt{2}}\right)}{A^{\alpha n}} - R(x) + \frac{T_{A^N}\left(\frac{x}{\sqrt{2}}\right)}{A^{\alpha N}} \right\|$$

$$-\frac{1}{A^{\alpha N}}\frac{1}{A^\alpha - 1}$$

We now notice that $T_{A^N}(x/\sqrt{2})$ is equal to 1 at least $(A^N/8) + 1$ times in [0, 1] and between each of those points it has a point where it is equal to -1. For

$$\sum_{n=1}^{N-1} \frac{T_{A^n}\left(\frac{x}{\sqrt{2}}\right)}{A^{\alpha n}} - R(x)$$

to have the same sign as $T_{A^N}(x/\sqrt{2})$ at each of those (one and minus one) points, it is necessary for

$$\sum_{n=1}^{N-1} \frac{T_{A^n}\left(\frac{x}{\sqrt{2}}\right)}{A^{\alpha n}} - R(x)$$

to have at least $A^N/4$ zeros. However,

$$\sum_{n=1}^{N-1} \frac{T_{A^n}\left(\frac{x}{\sqrt{2}}\right)}{A^{\alpha n}} - R(x)$$

can have at most A^{N-1} zeros, which is smaller than $A^N/4$ and consequently

$$\left\| \sum_{n=1}^{N-1} \frac{T_{A^n}\left(\frac{x}{\sqrt{2}}\right)}{A^{\alpha n}} - R(x) + \frac{T_{A^N}\left(\frac{x}{\sqrt{2}}\right)}{A^{\alpha N}} \right\| > \left\| \frac{T_{A^N}\left(\frac{x}{\sqrt{2}}\right)}{A^{\alpha N}} \right\| = \frac{1}{A^{\alpha N}}.$$

Therefore

$$\|f(x) - R(x)\| > \frac{1}{A^{\alpha N}} - \frac{1}{A^{\alpha N}}\frac{1}{A^\alpha - 1} = \frac{1}{A^{\alpha N}}\frac{A^\alpha - 2}{A^\alpha - 1}.$$

Since $R(x)$ was any member of $\mathscr{R}_{A^{N-1}}$ we must have

$$R_{A^{N-1}}(f) > \frac{1}{A^{\alpha N}}\frac{A^\alpha - 2}{A^\alpha - 1}.$$

Now let $A^{(N-1)} \leq n < A^N$. Then

$$R_n(f) \geq R_{A^N}(f) > \frac{1}{A^{\alpha(N+1)}}\frac{A^\alpha - 2}{A^\alpha - 1} = \frac{1}{A^{\alpha(N-1)}A^{2\alpha}}\frac{A^\alpha - 2}{A^\alpha - 1}$$

$$\geq \frac{1}{n^\alpha}\left[\frac{1}{A^{2\alpha}}\frac{A^\alpha - 2}{A^\alpha - 1}\right] = \frac{C(\alpha)}{n^\alpha}. \blacksquare$$

(2) Counter Examples

For polynomials we have found a bound on derivatives in terms of the functions. That is, $\|P'\| \leq n^2 \|P\|$ if $P \in \mathscr{P}_n$. Can we say anything similar for $\mathscr{R}_n$? This question is answered by the following counter example. Let $f(x) = 1/(1+Kx)$ $(K > 0)$. Then $f \in \mathscr{R}_1$ and $\|f\| = 1$. However, $\|f'\| = \|K/(1+Kx)^2\| = K$ and we can choose K arbitrarily large. Thus $\|f'\|/\|f\|$ is unbounded for $f \in \mathscr{R}_1 \subset \mathscr{R}_n$.

So far we have just talked about rational algebraic polynomials. However, we can prove the same sort of theorems for rational trigonometric polynomials. That is, the correct order of approximation by rational trigonometric polynomials of degree n (defined in the expected manner) to the class $\mathscr{S}^*$ is $1/n$ and to the class lip α is $1/n^\alpha$. We can also prove that we can't say anything about norms of derivatives in terms of norms of functions. Moreover, while for trigonometric polynomials we have inverse theorems, we do not have them for rational trigonometric polynomials.

THEOREM 3. *There is a function $f(x)$ such that $R_n^*(f) \leq C/n^\alpha$ but $f \notin \text{lip } \alpha$ $(0 < \alpha < 1)$.*

Proof. Let

$$f(x) = \sum_{n=1}^{\infty} \frac{1}{n^{10}} \frac{1}{1 + 2^n \sin^2 x}.$$

To prove we can approximate $f(x)$ by n-th degree trigonometric polynomials to within C/n^α, we consider $f(x)$ minus its N-th partial sum where $2N \leq n \leq 2(N+1)$. (We must remember that its N-th partial sum is a rational polynomial of degree $2N$.) Then

$$\left| f(x) - \sum_{K=1}^{N} \frac{1}{K^{10}} \frac{1}{1 + 2^K \sin^2 x} \right| = \left| \sum_{K=N+1}^{\infty} \frac{1}{K^{10}} \frac{1}{1 + 2^K \sin^2 x} \right|$$

$$= \sum_{K=N+1}^{\infty} \left| \frac{1}{K^{10}} \frac{1}{1 + 2^K \sin^2 x} \right|$$

$$\leq \sum_{K=N+1}^{\infty} \frac{1}{K^{10}}$$

$$\leq \int_N^{\infty} \frac{dx}{x^{10}} = \frac{1}{9N^9}$$

$$\leq \frac{4^9}{9} \frac{1}{n^9} \quad \text{(for } n \geq 4\text{)}$$

$$\leq \frac{4^9}{9} \frac{1}{n^\alpha}$$

for any $\alpha \in (0, 1)$. On the other hand, to prove that $f \notin \text{lip } \alpha$ for any $\alpha \in (0, 1)$ it suffices to prove that

$$\frac{|f(h) - f(0)|}{|h|^\alpha}$$

is unbounded for $h = 1/2^{j/2}$, $j = 1, 2, 3, \cdots$. We have

$$|f(h) - f(0)| = \left| \sum_{K=1}^{\infty} \frac{1}{K^{10}} \left(\frac{1}{1 + 2^K \sin^2 h} - 1 \right) \right|$$

$$= \sum_{K=1}^{\infty} \frac{1}{K^{10}} \frac{2^K \sin^2 h}{1 + 2^K \sin^2 h}$$

$$\geq \sin^2 h \sum_{K=1}^{j} \frac{1}{K^{10}} \frac{2^K}{1 + 2^K h^2}$$

$$\geq \frac{\sin^2 h}{2} \sum_{K=1}^{j} \frac{2^K}{K^{10}}$$

$$\geq \frac{\sin^2 h}{2 j^{10}} \sum_{K=1}^{j} 2^K$$

$$= \frac{\sin^2 h}{2 j^{10}} (2^{j+1} - 2)$$

$$\geq \frac{\left(\frac{2h}{\pi}\right)^2 2^j}{2 j^{10}}$$

$$= \frac{2}{\pi^2 j^{10}}$$

Therefore,

$$\frac{|f(h) - f(0)|}{|h|^\alpha} = \frac{|f(2^{-j/2}) - f(0)|}{2^{-j\alpha/2}} \geq \frac{2(2^{j\alpha/2})}{\pi^2 j^{10}}$$

which blows up as j approaches infinity. ∎

(3) Rational Approximation to $|x|$

So far, from what we have seen, rational approximation is not much better than regular polynomial approximation. For whole classes of functions like lip α and $\mathscr{S}$, rational approximation provides no improvement and seemingly, the only time we would get better approximation is in approximating rational polynomials themselves. However, in [26] it is proved that there is a big improvement in approximating $|x|$, and subsequently all sorts of theorems were proven showing better approximation for classes of functions of which $|x|$ is just a particular example.

Note. In this section, the relevant interval is $[-1, 1]$ and $E_n(f)$ and $R_n(f)$ are changed to account for the different interval under consideration.

We have seen that there exist C_1 and C_2 such that

$$\frac{C_1}{n} \leq E_n(|x|) \leq \frac{C_2}{n}.$$

That is, the correct order of polynomial approximation to $|x|$ on $[-1, 1]$ is $1/n$. However, for rational polynomials, the order of approximation is much smaller as seen in the following

THEOREM 4.

$$\frac{1}{2} e^{-9\sqrt{n+1}} \leq R_{n+1}(|x|) \leq 3 e^{-\sqrt{n}}.$$

Proof. We will first prove $R_{n+1}(|x|) \leq 3 e^{-\sqrt{n}}$. Let $\xi = e^{-1/\sqrt{n}}$ and let $P(x) = (x + 1)(x + \xi) (\cdots)(x + \xi^{n-1})$. Set

$$Q(x) = x \left[\frac{P(x) - P(-x)}{P(x) + P(-x)} \right].$$

Then $Q(x) \in \mathscr{R}_{n+1}$ and to prove $R_{n+1}(|x|) \leq 3 e^{-\sqrt{n}}$, it obviously suffices to prove $||x| - Q(x)| \leq 3 e^{-\sqrt{n}}$ for all $x \in [-1, 1]$. Since both $|x|$ and $Q(x)$ are even functions, it obviously suffices to restrict x to $[0, 1]$. On $[0, e^{-\sqrt{n}}]$, $P(x)$ and $P(-x)$ are positive and so

$$\left| \frac{P(x) - P(-x)}{P(x) + P(-x)} \right| \leq 1$$

Thus $|Q(x)| \leq |x|$ and $||x| - Q(x)| \leq 2|x| \leq 2 e^{-\sqrt{n}} < 3 e^{-\sqrt{n}}$. We are thus left to consider $||x| - Q(x)|$ on $[e^{-\sqrt{n}}, 1]$. Since

$$e^{-\sqrt{n}} = \xi^n < \xi^{n-1} < \; = \; < \xi < \xi^0 = 1$$

we can restrict x to $[\xi^{j+1}, \xi^j]$ where $j = 0, \ldots, n - 1$. There we have

$$\left| \frac{P(-x)}{P(x)} \right| = \left(\frac{1-x}{1+x} \right) \left(\frac{\xi - x}{\xi + x} \right) \cdots \left(\frac{\xi^j - x}{\xi^j + x} \right) \left(\frac{x - \xi^{j+1}}{x + \xi^{j+1}} \right) \cdots \left(\frac{x - \xi^{n-1}}{x + \xi^{n-1}} \right)$$

$$\leq \left(\frac{1 - \xi^n}{1 + \xi^n} \right) \left(\frac{\xi - \xi^n}{\xi + \xi^n} \right) \cdots \left(\frac{\xi^j - \xi^n}{\xi^j + \xi^n} \right) \left(\frac{\xi^j - \xi^{j+1}}{\xi^j + \xi^{j+1}} \right) \cdots \left(\frac{\xi^j - \xi^{n-1}}{\xi^j + \xi^{n-1}} \right)$$

$$= \prod_{K=1}^{n} \left(\frac{1 - \xi^K}{1 + \xi^K} \right)$$

$$\le e^{-2[\xi+\xi^2+\cdots+\xi^n]}\left(\text{since } \frac{1-x}{1-x} \le e^{-2x} \text{ for } x \ge 0\right)$$

$$= e^{-2[(\xi(1-\xi^n)/(1-\xi)]} \le e^{-\sqrt{n}}$$

(we have just used the inequality $(2\xi(1-\xi^n))/(1-\xi) \ge \sqrt{n}$ for $n \ge 4$, keeping in mind that $\xi = e^{-(1/\sqrt{n})}$).

Keeping this estimate for $|(P(-x))/(P(x)|$ in mind, we have, for $x \in [e^{-\sqrt{n}}, 1]$,

$$||x| - Q(x)| = \left| x - x\frac{P(x) - P(-x)}{P(x) + P(-x)} \right|$$

$$= 2|x| \left| \frac{P(-x)}{P(x) + P(-x)} \right| = 2|x| \left| \frac{1}{\frac{P(x)}{P(-x)} + 1} \right|$$

$$\le \frac{2}{\left|\frac{P(x)}{P(-x)}\right| - 1} \le \frac{2}{e^{\sqrt{n}} - 1} \le 3 e^{-\sqrt{n}} \text{ for } n \ge 2.$$

Thus, we have proven, $R_{n+1}(|x|) \le 3 e^{-\sqrt{n}}$. To prove,

$$R_{n+1}(|x|) \ge \frac{1}{2} e^{-9\sqrt{n+1}}$$

we will need the following lemma which will be proven later.

LEMMA 5. (1) $\int_a^b \log \left|\frac{x+z}{x-z}\right| \frac{dx}{x} \ge -\frac{\pi^2}{2}$ if $0 \le a < b$.

(2) If $P \in \mathscr{P}_{2n}$, there exists $x \in [e^{-\sqrt{2n}}, 1]$ such that

$$x \left|\frac{P(-x)}{P(x)}\right| > e^{-6\sqrt{2n}}.$$

Now assume $R(x)$ is a rational polynomial of degree n such that

$$|||x| - R(x)|| \le \frac{1}{2} e^{-9\sqrt{n}}.$$

Let

$$R_1(x) = \frac{R(x) + R(-x)}{2} - R(0).$$

Then

$$||x| - R_1(x)| = \frac{1}{2}|(|x| - R(x)) + (|x| - R(-x)) + 2(0 - R(0))|$$

RATIONAL APPROXIMATION 79

$$\leq 2\||x| - R(x)\| \leq e^{-9\sqrt{n}} < e^{-6\sqrt{2n}}.$$

We leave as an exercise the proof that $R_1(x)$, as an even rational polynomial of degree less than or equal to $2n$ which vanishes at 0, can be written as

$$\frac{x^2 P_1(x^2)}{P_2(x^2)}$$

where $P_1 \in \mathscr{P}_{n-1}$, $P_2 \in \mathscr{P}_n$. Let $x \in [e^{-\sqrt{2n}}, 1]$. Then

$$e^{-6\sqrt{2n}} > \left|x - \frac{x^2 P_1(x^2)}{P_2(x^2)}\right| = \left|x\frac{P_2(x^2) - xP_1(x^2)}{P_2(x^2)}\right|$$

$$\geq \left|x\frac{P_2(x^2) - xP_1(x^2)}{P_2(x^2) + xP_1(x^2)}\right|$$

(since for $x_1 \in [e^{-\sqrt{2n}}, 1]$, the sign of $P_1(x_1^2)$ is the same as the sign of $P_2(x_1^2)$ because otherwise,

$$\left|x_1 - x_1^2\frac{P_1(x_1^2)}{P_2(x_1^2)}\right| \geq |x_1| \geq e^{-\sqrt{2n}} > e^{-6\sqrt{2n}}).$$

Letting $P(x) = P_2(x^2) - xP_1(x^2)$, we have $P \in \mathscr{P}_{2n}$ and for

$$x \in [e^{-\sqrt{2n}}, 1], \left|x\frac{P(x)}{P(-x)}\right| \leq e^{-6\sqrt{2n}}$$

which contradicts parts (2) of the lemma. Thus, once we have proven the lemma, we'll have proven the non-existence of $R(x) \in \mathscr{R}_n$ such that

$$\||x| - R(x)\| \leq \frac{1}{2} e^{-9\sqrt{n}}$$

or that $R_n(|x|) > (1/2)e^{-9\sqrt{n}}$.

Proof of lemma 5. (1) Let $z = U + iV$. Then

$$\left|\frac{x+z}{x-z}\right| = \sqrt{\frac{(x+U)^2 + V^2}{(x+U)^2 + V^2}} \geq \sqrt{\frac{(x-|U|)^2 + V^2}{(x+|U|)^2 + V^2}}$$

$$\geq \sqrt{\frac{(x-|U|)^2}{(x+|U|)^2}} = \left|\frac{x-|U|}{x+|U|}\right|$$

Therefore

$$\int_a^b \log\left|\frac{x+z}{x-z}\right|\frac{dx}{x} \geq \int_a^b \log\left|\frac{x-|U|}{x+|U|}\right|\frac{dx}{x}$$

$$= \int_{a/|U|}^{b/|U|} \log\left|\frac{x-1}{x+1}\right|\frac{dx}{x}$$

$$\geq \int_0^\infty \log\left|\frac{x-1}{x+1}\right|\frac{dx}{x}$$

$$= 2\int_0^1 \log\left|\frac{x-1}{x+1}\right|\frac{dx}{x}$$

$$= 2\int_0^1 -2\left(1 + \frac{x^2}{3} + \frac{x^4}{5} + \cdots\right)dx$$

$$= -4\sum_{K=0}^\infty \frac{1}{(2K+1)^2} = \frac{-\pi^2}{2}.$$

To prove part (2), we let $P(x) = C\prod_{i=1}^{2n}(x - z_i)$. Then

$$\int_{e^{-\sqrt{2n}}}^1 \log\left(x\left|\frac{P(-x)}{P(x)}\right|\right)\frac{dx}{x} = \int_{e^{-\sqrt{2n}}}^1 \left(\log x + \sum_{i=1}^{2n} \log\left|\frac{x+z_i}{x-z_i}\right|\right)\frac{dx}{x}$$

$$= -n + \sum_{i=1}^{2n}\int_{e^{-\sqrt{2n}}}^1 \log\left|\frac{x+z_i}{x-z_i}\right|\frac{dx}{x}$$

$$\geq -n - \frac{2n\pi^2}{2} > -12n.$$

Now, if

$$\left|x\frac{P(-x)}{P(x)}\right| \leq e^{-6\sqrt{2n}}$$

in $[e^{-\sqrt{2n}}, 1]$, then

$$\int_{e^{-\sqrt{2n}}}^1 \log\left|x\frac{P(-x)}{P(x)}\right|\frac{dx}{x} \leq -6\sqrt{2n}\int_{e^{-\sqrt{2n}}}^1 \frac{dx}{x} = -12n$$

which is a contradiction and the lemma is proven. ∎

CHAPTER VIII

GENERALIZED POLYNOMIALS

So far, we have considered the approximation of various function classes by polynomials (or trigonometric polynomials). We will now try to generalize this concept somewhat.

Both polynomials $\varepsilon \mathscr{P}_n$ and trigonometric polynomials $\varepsilon \mathscr{T}_n$ are examples of elements of finite dimensional vector spaces. For our generalization, we take a real normed linear space B and a finite dimensional subspace $\phi \subseteq B$. That is, $\phi = \text{sp}\{\varphi_1, \varphi_2, \ldots, \varphi_n\}$ where $\varphi_i \in B$, $\{\varphi_i\}_{i=1}^n$ is linearly independent and $\text{sp}\{\varphi_1, \ldots, \varphi_n\}$ means the span of $\{\varphi_1 \varphi_2, \ldots, \varphi_n\}$ or the set of all linear combinations of $\{\varphi_i\}_{i=1}^n$.

For any $f \in B$ we consider $\inf \|f - a_1\varphi_1 \cdots a_n\varphi_n\|$ (where $\|\cdot\|$ is the norm of B) and we call this $E_\phi(f)$. This $E_\phi(f)$ measures how close f is to the nearest element of ϕ, the same way that $E_n(f)$ measures how close f is to the nearest element of $\mathscr{P}_n$. As a matter of fact, $E_n(f)$ is the same thing as $E_{\mathscr{P}_n}(f)$.

(1) Best Approximation

The following theorem, a generalization of theorems 3.1 and 3.6, states the existence of an "polynomial" of best approximation.

THEOREM 1. For each $f \in B$, there exists a $\varphi \in \phi$ such that $\|f - \varphi\| = E_\phi(f)$; i.e., there exist $a_1, a_2, \ldots, a_n$ such that

$$\|f - a_1\varphi_1 - \cdots - a_n\varphi_n\| = E_\phi(f).$$

Proof. Let Ψ_k be such that $\Psi_k \in \phi$ and $\lim_{k \to \infty} \|\Psi_k - f\| = E_\phi(f)$.

Let $\|\cdot\|_1$ be defined on ϕ as the maximum coefficients norm, *i.e.*, $\|a_1\varphi_1 + \cdots + a_n\varphi_n\|_1 = \max|a_j|$. Then, as we proved before, we can find A and B such that $\|\varphi\|_1 \leq A\|\varphi\| \leq B\|\varphi\|_1$ for all $\varphi \in \phi$.

Since $\|f - \Psi_k\|$ converges to $E_\phi(f)$ we know there is an M such that $\|f - \Psi_k\| \leq M$. Consequently,

$$\|\Psi_k\|_1 \leq A\|\Psi_k\| \leq A[\|\Psi_k - f\| + \|f\|] \leq A[M + \|f\|] = M_1.$$

By writing Ψ_k as $a_{1,k}\varphi_1 + \cdots + a_{n,k}\varphi_n$ we can interpret $\|\Psi_k\|_1 \leq M_1$ as $\sup_{\substack{i=1,\ldots,n \\ k=1,2,\ldots}} |a_{i,k}| \leq M_1$. By then considering, for $i = 1, 2, \ldots, n$, the sequence $\{a_{i,k}\}_{k=1}^\infty$ we see that each is a bounded sequence. By extracting subsequence after subsequence and relabeling we get $\lim_{k\to\infty} a_{i,k} = a_i$, $i = 1, 2, \ldots, n$.

Let $\varphi = a_1\varphi_1 + \cdots + a_n\varphi_n$. Then $\|\Psi_k - \varphi\|_1$ goes to 0 and consequently $\|\Psi_k - \varphi\|$ goes to 0. For a given $\varepsilon > 0$, by taking a large k, we can get $\|f - \Psi_k\| + \|\Psi_k - \varphi\| \leq E_\phi(f) + 2\varepsilon$. Then,

$$E_\phi(f) \leq \|f - \varphi\| \leq \|f - \Psi_k\| + \|\Psi_k - \varphi\| \leq E_\phi(f) + 2\varepsilon.$$

Since ε was arbitrary, we get $E_\phi(f) = \|f - \varphi\|$. ∎

Now that we have proven the existence of a "polynomial" of best approximation, a natural question to ask is that of uniqueness of the "polynomial" of best approximation. As we shall see, that depends on B and ϕ.

Definition. A normed linear space B is said to be strictly convex if f_1 and f_2 in B, $f_1 \neq f_2$, $\|f_1\| = \|f_2\|$ and $0 < \alpha < 1$ imply

$$\|\alpha f_1 + (1 - \alpha) f_2\| < \|f_1\|.$$

THEOREM 2. *In any strictly convex space B, the "polynomial" of best approximation is unique for every ϕ and every f.*

Proof. Assume that for some $\phi \subseteq B$ and $f \in B$, we can find $\varphi_1 \in \phi$ and $\varphi_2 \in \phi$ such that $\varphi_1 \neq \varphi_2$ and $\|f - \varphi_1\| = \|f - \varphi_2\| = E_\phi(f)$. Then

$$E_\phi(f) \leq \left\|f - \frac{1}{2}(\varphi_1 + \varphi_2)\right\| = \left\|\frac{1}{2}(f - \varphi_1) + \frac{1}{2}(f - \varphi_2)\right\|$$
$$< \|f - \varphi_1\| = E_\phi(f)$$

which is a contradiction.

Example 1. $L^p[a,b]$, where $1 < p < \infty$, is strictly convex. This is shown as follows: For these classes, Minkowski's inequality says that $\|f_1 + f_2\| \leq \|f_1\| + \|f_2\|$ with equality iff $f_1 = Kf_2$ where K is a positive constant. We are given f_1 and f_2, $f_1 \neq f_2$ and $\|f_1\| = \|f_2\|$. We take $\alpha \in (0,1)$. Then, by Minkowski's inequality,

$$\|\alpha f_1 + (1 - \alpha) f_2\| \leq \alpha\|f_1\| + (1 - \alpha)\|f_2\| = \|f_1\|$$

with equality iff $\alpha f_1 = K(1 - \alpha) f_2$ or $f_1 = (K(1 - \alpha)/\alpha) f_2$. However, since $\|f_1\| = \|f_2\|$ we get that $|K(1 - \alpha)/\alpha| = 1$ and since $K(1 - \alpha)/\alpha > 0$ we get that $K(1 - \alpha)/\alpha = 1$. However, since we also assume $f_1 \neq f_2$ we cannot have $K(1-\alpha)/\alpha = 1$ and thus couldn't have equality. Thus $\|\alpha f + (1-\alpha)f_2\| < \|f_1\|$ and we have proven strict convexity.

The spaces $C[a,b]$ and $L^1[a,b]$ are not strictly convex. Moreover, not only is the hypothesis of the previous theorem not true, but the conclusion is also. This is seen in the following examples.

Example 2. Consider $C[0,1]$ and $\phi = \text{sp}\{\varphi\}$ where $\varphi(t) = t$. If we wish to approximate $f(t) \equiv 1$ we see that $\|f - a\varphi\| = 1$ if $0 \leq a \leq 2$ and $\|f - a\varphi\| > 1$ if $a \notin [0,2]$. Thus, for any $a \in [0,2]$, $a\varphi$ is a best approximation to f.

Example 3. Consider $L^1[-1,1]$ and $\phi = \text{sp}\{\varphi\}$ where $\varphi(t) = \text{sgn } t$. If we wish to approximate

$$f(t) = \begin{cases} -2 & -1 \leq t \leq 0 \\ 1 & 0 < t < 1 \end{cases}$$

then we have that
(a) $a \leq 0 \Rightarrow \int_{-1}^{1} |f(t) - a\varphi(t)|\, dt \geq 3$.
(b) $0 \leq a \leq 1 \Rightarrow \int_{-1}^{1} |f(t) - a\varphi(t)|\, dt = 3 - 2a$.
(c) $1 \leq a \leq 2 \Rightarrow \int_{-1}^{1} |f(t) - a\varphi(t)|\, dt = 1$.
(d) $a \geq 2 \Rightarrow \int_{-1}^{1} |f(t) - a\varphi(t)|\, dt = 2a - 3$.
Thus, any $a \in [1,2]$ will give us a best approximation.

(2) Lower Bounds for $E_\phi(\mathscr{S})$

The next extension of polynomial approximation that we consider, is the question of how well a space of "polynomials" can approximate the class $\mathscr{S}$. (By $\mathscr{S}$ we now mean the class of real valued functions f defined on the compact metric space (M,ρ) which satisfy

$$|f(x) - f(y)| \leq \rho(x,y)$$

for all $x,y \in M$. When there is likelihood of confusion as to which metric space we are referring to, we will write this as $\mathscr{S}(M)$.)

As before, we let $E_\phi(\mathscr{S}) = \sup_{f \in \mathscr{S}} E_\phi(f)$.

We will devlop a method of getting a lower bound on $E_\phi(\mathscr{S})$ which will depend on the geometry of (M,ρ) (and surprisingly, not on ϕ). First, though, we will need the concept of an open orthant in R^n which is just a generalization of an open quadrant in R^2.

Definition. An open orthant in R^n is an n-tuple of $+$ and $-$ signs; i.e., something of the form $(+,-,-,+,\ldots,+)$.

Example. The first quadrant in R^2 is $(+,+)$.

LEMMA 3. Let T be a proper subspace of R^n. Then T misses some open orthant. That is, there is some n-tuple of $+$ and $-$ signs which is never found in T.

Proof. Assume that the theorem is false and we'll show that we can find all the standard basis vectors in T. That, of course, would contradict T being a proper subspace. We'll demonstrate a method for proving that $(1, 0, 0, \ldots, 0) \in T$ and it will be obvious we could just as easily show that $(0, 0, \ldots, 0, 1, 0, \ldots, 0) \in T$.

By assumption, T contains a vector $(a_1, a_2, \ldots, a_n)$ with $a_i > 0$, and a vector $(b_1, b_2, \ldots, b_n)$ with all $b_i > 0$ except $b_2 < 0$. Then T would contain $-b_2(a_1, a_2, \ldots, a_n) + a_2(b_1, b_2, \ldots, b_n) = (c_1, 0, c_3, c_3, \ldots, c_n)$, which is of the form $(+, 0, +, +, \ldots, +)$. Similary T would contain a $(d_1, 0, d_3, d_4, \ldots, d_n)$ of the form $(+, 0, -, +, +, \ldots, +)$. Then T would contain $-d_3(c_1, 0, c_3, \ldots, c_n) + c_3(d_1, 0, \ldots, d_n)$ which is of the form $(+, 0, 0, +, +, \ldots, +)$. Continuing in this manner we could get something of the form $(+, 0, 0, \ldots, 0)$ or $(1, 0, 0, \ldots, 0)$. By our previous comments we have proven the lemma. ∎

Before we consider a general real normed linear space B, we restrict our attention to the space $C[0, 1]$ (of real-valued continuous functions on $[0, 1]$). We saw before that if $\phi = \mathscr{P}_{n-1}$ then $E_\phi(\mathscr{S}) = E_{n-1}(\mathscr{S}) \geq 1/2n$. We will now prove that the same lower bound holds for any n dimensional ϕ.

THEOREM 4. If ϕ is an n-dimensional subspace of $C[0, 1]$, then

$$E_\phi(\mathscr{S}) \geq \frac{1}{2n}.$$

Proof. For each $\varphi \in \phi$, consider the mapping given by

$$\varphi \to \left(\varphi(0), \varphi\left(\frac{1}{n}\right), \varphi\left(\frac{2}{n}\right), \ldots, \varphi(1)\right)$$

of $\phi \to R^{n+1}$. The image space is an n-dimensional subspace of R^{n+1}; i.e., a proper subspace of R^{n+1}. By the lemma, there is an open orthant of R^{n+1} which is never intersected by the image space. Define $f(k/n)$ to be $1/2n$ if there is a $+$ sign in the $k + 1$-st coordinate of that orthant and $-(1/2n)$ if there is a $-$ sign. Define $f(x)$ to be linear in between. It is obvious that $f \in \mathscr{S}$. Moreover, for any $\varphi \in \phi$, $\|f - \varphi\| \geq 1/2n$ since at at least one of the points k/n, $\varphi(k/n)$ will have the opposite sign (or 0) to $f(k/n)$ because of the missing orthant. Thus $E_\phi(f) \geq 1/2n$ and certainly, $E_\phi(\mathscr{S}) \geq 1/2n$. ∎

Thus, we see that for any n-dimensional space ϕ, we can't do any better than $1/2n$. By Jackson's theorem, we saw that $E_{\mathscr{P}_{n-1}}(\mathscr{S}) \leq c/n$.

Thus, algebraic polynomials give us what is essentially the best possible approximation.

We also note, that in the proof we did not use the continuity of φ. The whole proof could have applied as well to $L^\infty[0, 1]$. That is, if ϕ is an n-dimensional subspace of $L^\infty[0, 1]$, then $E_\phi(\mathscr{S}) \geq 1/2n$. However, whereas before we could find a $\phi \subset C[0, 1]$ such that $E_\phi(\mathscr{S}) \leq c/n$, for $L^\infty[0, 1]$ we can actually get $1/2n$ which obviously is the best we can do. We do this as follows: For $k = 1, 2, \ldots, n - 1$, we let

$$\varphi_k(x) = \begin{cases} 1 & x \in \left[\frac{k-1}{n}, \frac{k}{n}\right) \\ 0 & \text{elsewhere} \end{cases}$$

while

$$\varphi_n(x) = \begin{cases} 1 & x \in \left[\frac{n-1}{n}, 1\right] \\ 0 & \text{elsewhere} \end{cases}.$$

Then ϕ is just the space of step functions with jumps at

$$\frac{1}{n}, \frac{2}{n}, \ldots, \frac{n-1}{n}.$$

For $f \in \mathscr{S}$, we approximate f by $a_1\varphi_1 + \cdots + a_n\varphi_n$ where

$$a_k = \frac{1}{2}[\max f + \min f]$$

where $\max f$ and $\min f$ are taken with respect to $[(k-1)/n, k/n]$. The error in each subinterval is less than $1/2[\max f - \min f]$ which, since $f \in \mathscr{S}$, is less than $1/2n$. Thus $E_\phi(f) \leq 1/2n$ and therefore $E_\phi(\mathscr{S}) \leq 1/2n$.

In order to give a lower bound for a more general B, we need the following:

Definition. For (M, ρ) a compact metric space, we define

$$\varepsilon_n = \varepsilon_n(M) = \max_{x_1, x_2, \ldots, x_{n+1} \in M} \min_{i \neq j} \rho(x_i, x_j)$$

Thus

$$\varepsilon_1 = \max_{x_1, x_2 \in M} \rho(x_1, x_2)$$

is the usual diameter of M.

Example 1. Let $M = [0, 1]$ with ρ the usual metric. Then, if

$$x_k = \frac{k-1}{n}, \quad k = 1, 2, \ldots, n+1$$

we could get $\varepsilon_n \geq 1/n$. On the other hand, by considering the n intervals $[k/n, (k+1)/n]$, $k = 0, \ldots, n-1$, we see that any time we have $n+1$ points we must have two of them in the same interval. Thus $\varepsilon_n \leq 1/n$ and consequently $\varepsilon_n = 1/n$.

In general, it is fairly hard to calculate $\varepsilon_n(M)$ exactly. What we will do, frequently, is either just get a lower bound on $\varepsilon_n(M)$ or else calculate the asymptotic behaviour of $\varepsilon_n(M)$ as n approaches infinity. For instance, it can be shown that for M the unit disc, $\varepsilon_n(M)$ behaves like $1/\sqrt{n}$.

In terms of this ε_n we will now get a lower bound for the Banach space B of continuous real valued functions on the compact metric space (M, ρ), with the norm of B being the sup norm. (Actually, we will even stretch B to include "step" functions.)

THEOREM 5. Let ϕ be n-dimensional. Then

$$E_\phi(\mathscr{S}) \geq \frac{\varepsilon_n(M)}{2}$$

Proof. We will generalize the proof of theorem 4. Let $x_1, x_2, \ldots, x_{n+1}$ give us an extremal configuration. That is, $\rho(x_i, x_j) \geq \varepsilon_n(M)$ for $1 \leq i < j \leq n+1$. For each $\varphi \in \phi$, consider the mapping of $\phi \to R^{n+1}$ given by $\varphi \to (\varphi(x_1), \varphi(x_2), \ldots, \varphi(x_{n+1}))$. The image space is an n-dimensional subspace of R^{n+1}. By lemma 3, we know there is an open orthant of R^{n+1} which is never intersected by the image space. Pick out one such orthant and let T consist of those points x_j for which the j-th sign in the orthant is $-$ (as an example, if $(+, -, +, -, \ldots)$ is the orthant, then T consists of $\{x_2, x_4, x_6, \ldots\}$). Now let

$$f(x) = \rho(x, T) - \frac{\varepsilon_n(M)}{2} = \min_{x_i \in T} \rho(x, x_i) - \frac{\varepsilon_n(M)}{2}.$$

Consider any $x, y \in M$. We can find $x', y' \in T$ such that $\rho(x, x') = \rho(x, T)$ and $\rho(y, y') = \rho(y, T)$. Then $|f(x) - f(y)| = |\rho(x, x') - \rho(y, y')|$. However, $\rho(x, x') \leq \rho(x, y') \leq \rho(x, y) + \rho(y, y')$ and

$$\rho(y, y') \leq \rho(y, x') \leq \rho(y, x) + \rho(x, x').$$

Thus $|\rho(x, x') - \rho(y, y')| \leq \rho(x, y)$ and consequently

$$|f(x) - f(y)| \leq \rho(x, y)$$

and $f \in \mathscr{S}$.

Now take any $\varphi \in \phi$. Since $(\varphi(x_1), \ldots, \varphi(x_{n+1}))$ is not our orthant, we must have either:
(a) an $x_i \in T$ such that $\varphi(x_i) \geq 0$ or
(b) an $x_i \notin T$ such that $\varphi(x_i) \leq 0$.
In case (a) $f(x_i) - \varphi(x_i) = -(1/2)\varepsilon_n(M) - \varphi(x_i) \leq -(1/2)\varepsilon_n(M)$. In case (b) $f(x_i) - \varphi(x_i) \geq \varepsilon_n(M) - (1/2)\varepsilon_n(M) = (1/2)\varepsilon_n(M)$. Thus in either case $\|f - \varphi\| \geq |f(x_i) - \varphi(x_i)| \geq (1/2)\varepsilon_n(M)$ and we have $E_\phi(f) \geq (1/2)\varepsilon_n(M)$. Since we have proven that $f \in \mathscr{S}$, we have

$$E_\phi(\mathscr{S}) \geq \frac{1}{2}\varepsilon_n(M).\ \blacksquare$$

We note that when $M = [0, 1]$ the lower bound is the same as before.

Before, when we were studying approximation by algebraic polynomials, we saw that we could get an estimate on $E_n(f)$ in terms of $E_n(\mathscr{S})$. We shall see that we can get the same kind of estimate even in the more general case. First, though, we will need the following:

LEMMA 6. *Let (M, ρ) be a compact metric space and let $f(x)$ be in $\mathscr{S}(M_1)$ where $M_1 \subset M$. Then $f(x)$ can be extended to a function in $\mathscr{S}(M)$.*

Proof. By the principle of transfinite induction, we need only prove that we can extend $f(x)$ one point at a time.

Take a point $x' \in M - M_1$. If we define $f(x') = \alpha$, we have to make sure that $|f(x) - \alpha| \leq \rho(x, x')$ for all $x \in M_1$. This, of course, is equivalent to $-\rho(x', x) + f(x) \leq \alpha \leq \rho(x', x) + f(x)$ for all $x \in M_1$. That is the same as $\alpha \geq -\rho(x', y) + f(y)$ for $y \in M_1$ and $\alpha \leq \rho(x', x) + f(x)$ for $x \in M_1$. Thus, we can find such an α iff

$$-\rho(x', y) + f(y) \leq \rho(x', x) + f(x) \quad \text{for all } x, y \in M_1,$$

or

$$f(y) - f(x) \leq \rho(x', x) + \rho(x', y) \quad \text{for } x, y \in M_1.$$

However, for $x, y \in M_1$,

$$f(y) - f(x) \leq |f(x) - f(y)| \leq \rho(x, y) \leq \rho(x, x') + \rho(x', y).$$

Thus we can find an α and can extend f to $\mathscr{S}(M)$. $\blacksquare$

For a function f defined on (M, ρ) the modulus of continuity of f is defined in the obvious manner. That is,

$$\omega(f, \delta) = \sup_{\rho(x,y) \leq \delta} |f(x) - f(y)|$$

In terms of $\omega(f, \cdot)$ and $E_\phi(\mathscr{S})$, we can now get the promised estimate on $E_\phi(f)$.

POLYNOMIAL APPROXIMATION

Theorem 7. $E_\phi(f) \leq 5\omega(f, E_\phi(\mathscr{S}))$.

Proof. Let $S(x)$ denote a closed ball of radius $E_\phi(\mathscr{S})$ centered at x. We will form a set $M_1 \subset M$ as follows: Take any $x_1 \in M$. Then take any $x_2 \in M - S(x_1)$. Inductively, after taking $x_1, x_2, \ldots, x_n$ we take

$$x_{n+1} \in M - \bigcup_{i=1}^{n} S(x_i)$$

and we let $M_1 = \{x_1, x_2, \ldots\}$. We note that since M is totally bounded, M_1 is finite. We also note that $\rho(x_i, x_j) \geq E_\phi(\mathscr{S})$ for $i \neq j$.

Take any $f(x)$ defined on M. Then

$$|f(x_i) - f(x_j)| \leq \omega(f, \rho(x_i, x_j))$$
$$\leq \left(\frac{\rho(x_i, x_j)}{E_\phi(\mathscr{S})} + 1\right)\omega(f, E_\phi(\mathscr{S}))$$

(by lemma 4.7)

$$\leq 2\frac{\rho(x_i, x_j)}{E_\phi(\mathscr{S})}\omega(f, E_\phi(\mathscr{S})).$$

Thus

$$\frac{E_\phi(\mathscr{S})}{2\omega(f, E_\phi(\mathscr{S}))} f(x) \in \mathscr{S}(M_1).$$

By the lemma, we can extend it to a function $g \in \mathscr{S}(M)$. Then we can find a $\varphi \in \phi$ such that $\|g - \varphi\| \leq E_\phi(\mathscr{S})$. (We recall that $\|g - \varphi\| = \sup_{x \in M} |g(x) - \varphi(x)|$.)

Take any $x \in M$ and assume $x \in S(x_k)$. Then

$$\left|g(x) - \frac{E_\phi(\mathscr{S})f(x)}{2\omega(f, E_\phi(\mathscr{S}))}\right| \leq |g(x) - g(x_k)| + \left|g(x_k) - \frac{E_\phi(\mathscr{S})f(x_k)}{2\omega(f, E_\phi(\mathscr{S}))}\right|$$
$$+ \left|\frac{E_\phi(\mathscr{S})[f(x_k) - f(x)]}{2\omega(f, E_\phi(\mathscr{S}))}\right|$$
$$\leq \rho(x, x_k) + 0 + \frac{E_\phi(\mathscr{S})}{2\omega(f, E_\phi(\mathscr{S}))}\omega(f, \rho(x_k, x))$$
$$\leq E_\phi(\mathscr{S}) + 0 + \frac{E_\phi(\mathscr{S})}{2}$$
$$= \frac{3}{2} E_\phi(\mathscr{S}).$$

Thus, for any $x \in M$,

$$\left| \frac{E_\phi(\mathscr{S})f(x)}{2\omega(f, E_\phi(\mathscr{S}))} - \varphi(x) \right| \leq \left| \frac{E_\phi(\mathscr{S})f(x)}{2\omega(f, E_\phi(\mathscr{S}))} - g(x) \right| + |g(x) - \varphi(x)|$$

$$\leq \frac{5}{2} E_\phi(\mathscr{S}).$$

Therefore,

$$\left| f(x) - \frac{2\omega(f, E_\phi(\mathscr{S}))}{E_\phi(\mathscr{S})} \varphi(x) \right| \leq 5\omega(f, E_\phi(\mathscr{S}))$$

and since

$$\frac{2\omega(f, E_\phi(\mathscr{S}))}{E_\phi(\mathscr{S})} \varphi(x) \in \phi,$$

we have $E_\phi(f) \leq 5\omega(f, E_\phi(\mathscr{S}))$. ∎

(3) Approximation by Step Functions

Previously, when we were talking about $M = [0, 1]$, we saw that we had a lower bound of $1/2n$ which we were able to achieve with step functions.

For a more general M, when we have a lower bound $\varepsilon_n(M)/2$, how close can we come with step functions? (By a step function, we mean a finite linear combination of characteristic functions.)

The next theorem will prove that step functions can "essentially" achieve the lower bound; that is they can always come within twice the lower bound.

THEOREM 8. *For any n, there exists an n-dimensional space ϕ, of step functions, on the compact metric space (M, ρ) such that $E_\phi(\mathscr{S}) \leq \varepsilon_n(M)$.*

Proof. By compactness, we know there exists at least one $n + 1$ tuple $(x_1, x_2, \ldots, x_{n+1})$ such that $x_j \in M$ and $\rho(x_i, x_j) \geq \varepsilon_n(M)$ for $i \neq j$. Moreover, for each such $n + 1$ tuple we can find k_1, k_2 such that

$$\rho(x_{k_1}, x_{k_2}) = \varepsilon_n(M).$$

From all such $n + 1$ tuples pick a $(y_1, \ldots, y_{n+1})$ which is minimal in the sense that it has the fewest number of pairs achieving $\varepsilon_n(M)$. We assume $\rho(y_n, y_{n+1}) = \varepsilon_n(M)$.

We now claim that, if we take the union of closed balls of radius $\varepsilon_n(M)$ centered at $y_i, i = 1, \ldots, n$, then we include all of M. If not, then we can find a $z \in M$ such that $\rho(z, y_i) > \varepsilon_n(M), i = 1, 2, \ldots, n$. We would then have that the $n + 1$ tuple $(y_1, y_2, \ldots, y_n, z)$ is an $n + 1$

tuple considered earlier which would have fewer pairs achieving $\varepsilon_n(M)$ than $(y_1, y_2, \ldots, y_{n+1})$, contradicting the minimality of $(y_1, y_2, \ldots, y_{n+1})$.

We let S_1 be the closed ball of radius $\varepsilon_n(M)$ centered at y_1. We let S_2 be the intersection of $M - S_1$ and the closed ball of radius $\varepsilon_n(M)$ centered at y_2. In a similar manner we define $S_3, \ldots, S_n$. We then have $\bigcup_{i=1}^n S_i = M$ and $S_i \cap S_j$ is empty for $i \neq j$. We now let $\varphi_i(x)$ be the characteristic function of S_i (that is $\varphi_i(x) = 1$ if $x \in S_i$ and 0 if $x \notin S_i$) and let $\phi = \text{sp}\{\varphi_1, \varphi_2, \ldots, \varphi_n\}$.

We will now prove that $E_\phi(\mathscr{S}) \leq \varepsilon_n(M)$. Take any $f \in \mathscr{S}$. We wish to approximate f by $a_1\varphi_1 + \cdots + a_n\varphi_n$ which is a_i for $x \in S_i$. We let $a_i = f(y_i)$. Then

$$\|f - \varphi\| = \max_{i=1,\ldots,n} \sup_{x \in S_i} |f(x) - \varphi(x)|$$
$$= \max_{i=1,\ldots,n} \sup_{x \in S_i} |f(x) - f(y_i)| \leq \max_{i=1,\ldots,n} \sup_{x \in S_i} \rho(x, y_i)$$
$$\leq \varepsilon_n(M) .$$

Thus $E_\phi(f) \leq \varepsilon_n(M)$ and consequently $E_\phi(\mathscr{S}) \leq \varepsilon_n(M)$. ∎

We now are in the following situation. We know that if $M = [0, 1]$, then step functions can achieve the lower bound. Thus, the lower bound is achieved and step functions form a best n-dimensional approximating space. For a general M, however, we know only that step functions come within twice the lower bound. Thus the following two questions are still unanswered:

(1) Can the lower bound (of $\varepsilon_n(M)/2$) be achieved?

(2) Do step functions form a best n-dimensional approximating space?

For $n > 2$ (and arbitrary M) the answers to these questions are still unknown. However, for $n = 2$ we shall see that the answer to the second question is "yes" while the answer to the first question depends on M.

Notation. If $T \subset M$, T^c will denote its complement, $d(T)$, its diameter and

$$T(x) = \begin{cases} 1 & x \in T \\ 0 & x \notin T \end{cases}$$

We will prove: there is a $T \subset M$ such that if $\phi = \text{sp}\{T, T^c\}$, then $E_\phi(\mathscr{S}) = \min E_\Psi(\mathscr{S})$ where the min is taken over all 2-dimensional Ψ. This, of course, would answer the second question and to answer the first question we will have to rely on theorem 11 which will tell us that if $\phi = \text{sup}\{T, T^c\}$ then $E_\phi(\mathscr{S}) = (1/2) \max(d(T), d(T^c))$ and then check to see when that is $(1/2)\varepsilon_n(M)$.

The next theorem will tell us that in looking for a best space we should include the constant function.

THEOREM 9. *Let* $\phi = \sup\{\varphi_1, \varphi_2\}$. *Then, if* $1 \in \phi$, $E_\phi(\mathscr{S}) \leq (1/2)d(M)$, *while, if* $1 \notin \phi$, $E_\phi(\mathscr{S}) = \infty$.

Proof. If $1 \in \phi$, for each $f \in \mathscr{S}$ we approximate f by

$$\frac{1}{2}[\max f + \min f]$$

(both taken over M). Since

$$\left\| f - \frac{1}{2}[\max f + \min f] \right\| \leq \frac{1}{2}d(M),$$

we get that $E_\phi(f) \leq (1/2)d(M)$ and consequently $E_\phi(\mathscr{S}) \leq (1/2)d(M)$. If, however, $1 \notin \phi$, then $E_\phi(1) > 0$ (by compactness) and, since $E_\phi(n) = nE_\phi(1)$ and $n \in \mathscr{S}$, we can make $E_\phi(\mathscr{S})$ arbitrarily large. ∎

Thus, in looking for a best 2-dimensional approximating space, we know that we have to include 1. Our search then is for the second function. We now make some simplifications.

First of all, since we have 1 to approximate with, we have only to approximate those $f \in \mathscr{S}$ such that $\|f\| \leq d(M)$ since, for any other $f \in \mathscr{S}$, we can consider only $f(x) - f(x_0)$ where x_0 is fixed in M.

Second, now that we only have to approximate bounded functions, we need only consider bounded functions for our second function. Without loss of generality we consider only functions of norm ≤ 1.

Notation. $\mathscr{S}_0 = \mathscr{S}_0(M) = \{f \in \mathscr{S}, \|f\| \leq d(M)\}$. $E_g(\mathscr{S}) = E_G(\mathscr{S})$ for $G = \text{sp}\{1, g\}$.

The next theorem provides a crucial inequality.

THEOREM 10. *Let* $g(x)$ *be defined on* M *and let* $x_1, x_2,$ *and* x_3 *be points of* M *such that* $g(x_1) < g(x_2) < g(x_3)$. *Then*

$$E_g(\mathscr{S}) \geq \frac{[g(x_3) - g(x_2)]\rho(x_1, x_2) + [g(x_2) - g(x_1)]\rho(x_2, x_3)}{2[g(x_3) - g(x_1)]}.$$

Proof. Let $M_1 = \{x_1, x_2, x_3\}$ and let $f(x) = \rho(x, x_2)$. It is easy to see that $f \in \mathscr{S}_0(M) \subseteq \mathscr{S}(M_1)$. Let $a + bg(x)$ be a best approximation to f on M_1 (from $\text{sp}\{1, g\}$), and let $\delta = \max_{i=1,2,3} |a + bg(x_i) - f(x_i)|$. It is easy to see that the following alternation scheme must occur:

$$a + bg(x_1) - f(x_1) = -\delta$$
$$a + bg(x_2) - f(x_2) = \delta$$

$$a + bg(x_3) - f(x_3) = -\delta$$

(essentially, the reason is that, otherwise, better a and b could be found). Solving these equations for δ, we get

$$\delta = \frac{[g(x_3) - g(x_2)]\rho(x_1, x_2) + [g(x_2) - g(x_1)]\rho(x_2, x_3)}{2[g(x_3) - g(x_1)]}$$

Thus

$$E_g(\mathscr{S}) \geq \inf_{a,b} \sup_{x \in M} |a + bg(x) - f(x)| \geq \inf_{a,b} \sup_{x \in M_1} |a + bg(x) - f(x)| = \delta. \quad \blacksquare$$

Our next theorem tells us exactly what the "error" is if the second function is a characteristic function.

THEOREM 11. *If* $T \subseteq M$, $E_T(\mathscr{S}) = (1/2) \max(d(T), d(T^c))$.

Proof. Since $T^c(x) = 1 - T(x)$, sp $\{1, T\} =$ sp $\{T, T^c\}$. Let $f \in \mathscr{S}_0$. We want to approximate f by $aT(x) + bT^c(x)$ (which is a if $x \in T$ and b if $x \in T^c$). Let

$$a = \frac{1}{2}\left[\sup_{x \in T} f(x) + \inf_{x \in T} f(x)\right]$$

and let

$$b = \frac{1}{2}\left[\sup_{x \in T^c} f(x) + \inf_{x \in T^c} f(x)\right].$$

Then, for $x \in T$,

$$|f(x) - a| = \left|\frac{1}{2}[f(x) - \sup_{x \in T} f(x)] + \frac{1}{2}[f(x) - \inf_{x \in T} f(x)]\right|.$$

The two summands never have the same sign and, since $f \in \mathscr{S}_0$, each is less than $(1/2)d(T)$. Similarly $|f(x) - b| \leq (1/2)d(T^c)$ for $x \in T^c$. Thus, since f is arbitrary, $E_T(\mathscr{S}) \leq (1/2) \max(d(T), d(T^c))$.

Now let $\varepsilon > 0$ be given and choose $x_1, x_2 \in T$ such that

$$\rho(x_1, x_2) > d(T) - \varepsilon$$

and let $f(x) = \rho(x, x_2)$. Then $f \in \mathscr{S}_0$. For any a and b,

$$\|f - aT - bT^c\| \geq \max_{x = x_1, x_2} |f(x) - aT(x) - bT^c(x)|$$
$$= \max[|a|, |\rho(x_1, x_2) - a|]$$
$$\geq \frac{1}{2}\rho(x_1, x_2)$$

$$> \frac{1}{2}(d(T) - \varepsilon).$$

Thus $E_T(\mathscr{S}) \geq (1/2)d(T) - \varepsilon/2$. Similarly, $E_T(\mathscr{S}) \geq (1/2)d(T^c) - \varepsilon/2$. Since ε was arbitrary, we get $E_T(\mathscr{S}) \geq (1/2) \max(d(T), d(T^c))$ and thus equality. ∎

So far, we have a lower bound on the "error" with an arbitrary function and the exact "error" with a characteristic function. In the next theorem we will combine these two to prove that for certain functions, we can always do better with characteristic functions.

THEOREM 12. Le $g(x)$ have only a finite number of values. Then there is a $T \subseteq M$ such that $E_T(\mathscr{S}) \leq E_g(\mathscr{S})$.

Proof. Let $y_1, y_2, \ldots y_n$ (where $-1 \leq y_1 < y_2 < \cdots < y_n \leq 1$) be the values of $g(x)$. By the argument used in the second part of the proof of the previous theorem, $d(g^{-1}(y_k)) \leq 2E_g(\mathscr{S})$, $k = 1, \ldots, n$. Let $\cup_1 = g^{-1}(y_1)$ and let $\cup_k$ ($k > 1$) be defined inductively by

$$\cup_k = \cup_{k-1} \cup \{x \in g^{-1}(y_k): \quad \rho(x, x') \leq 2E_g(\mathscr{S}) \text{ for all } x' \in \cup_{k-1}\}.$$

It is obvious that $d(\cup_k) \leq 2E_g(\mathscr{S})$. If we set $T = \cup_n$, then, by the previous theorem, to prove $E_T(\mathscr{S}) \leq 2E_g(\mathscr{S})$ we just have to prove $d(T^c) \leq 2E_g(\mathscr{S})$.

If we assume that $d(T^c) > 2E_g(\mathscr{S})$ then we can find $x_1, x_2 \in T^c$ with $\rho(x_1, x_2) > 2E_g(\mathscr{S})$. For some j, k, $x_1 \in g^{-1}(y_j)$ and $x_2 \in g^{-1}(y_k)$ and since $d(g^{-1}(y_j)) \leq 2E_g(\mathscr{S})$ we must have $j \neq k$; we can assume $j < k$. Moreover since $g^{-1}(y_1) \subseteq T$ we must have $1 < j < k$. As $x_1 \in T^c$, there is an $x_0 \in g^{-1}(y_i)$ with $i < j$ and $\rho(x_0, x_1) > 2E_g(\mathscr{S})$. By theorem 10, since $g(x_0) < g(x_1) < g(x_2)$, we have

$$E_g(\mathscr{S}) \geq \frac{[g(x_2) - g(x_1)]\rho(x_1, x_0) + [g(x_1) - g(x_0)]\rho(x_2, x_1)}{2[g(x_2) - g(x_0)]}$$

$$> \frac{[g(x_2) - g(x_1)]2E_g(\mathscr{S}) + [g(x_1) - g(x_0)]2E_g(\mathscr{S})}{2[g(x_2) - g(x_0)]}$$

$$= E_g(\mathscr{S})$$

which is a contradiction. Therefore, $d(T^c) \leq 2E_g(\mathscr{S})$ and

$$E_T(\mathscr{S}) \leq E_g(\mathscr{S}). \blacksquare$$

What we have proven until now is that for a g which has only a finite number of values, we can find a $T(x)$ which is better. However, any function can approximated arbitrarily closely by such a $g(x)$. The next lemma proves that in approximating one function by another, the "error" doesn't change much.

LEMMA 13. Let $g(x)$ be non-constant and let $\|g_n - g\|$ converge to 0. Then $E_{g_n}(\mathscr{S})$ converges to $E_g(\mathscr{S})$.

Proof. In approximating $f \in \mathscr{S}_0$ by $a + bg(x)$ we can always assume $\|a + bg - f\| \le \|f\|$ (since otherwise we could do better with $a = b = 0$).

For every a, b let $L(a + bg) = b$. Then, since g is non-constant, L is a well defined linear functional on a finite dimensional vector space and is, therefore, bounded; *i.e.*, there exists a $c(g)$ such that

$$|b| = |L(a + bg)| \le c(g)\|a + bg\|.$$

Let a and b be arbitrary real numbers and let $f \in \mathscr{S}_0$. Then

$$\big|\|a + bg - f\| - \|a + bg_n - f\|\big| \le \|(a + bg - f) - (a + bg_n - f)\|$$
$$= |b|\|g - g_n\|$$
$$\le c(g)\|a + bg\|\|g - g_n\|$$
$$\le c(g)(\|a + bg - f\| + \|f\|)\|g - g_n\|$$
$$\le 2c(g)\|f\|\|g - g\|$$
$$\le 2d(M)c(g)\|g - g_n\|.$$

Thus,

$$\|a + bg - f\| \le \|a + bg_n - f\| + 2d(M)c(g)\|g - g_n\|.$$

Therefore,

$$\sup_{f \in \mathscr{S}_0} \inf_{a,b} \|a + bg - f\| \le \sup_{f \in \mathscr{S}_0} \inf_{a,b} \|a + bg_n - f\| + 2d(M)c(g)\|g - g_n\|$$

or

$$E_g(\mathscr{S}) \le E_{g_n}(\mathscr{S}) + 2d(M)c(g)\|g - g_n\|.$$

Similarly, starting with

$$\|a + bg_n - f\| \le \|a + bg - f\| + 2d(M)c(g)\|g - g_n\|,$$

we could get

$$E_{g_n}(\mathscr{S}) \le E_g(\mathscr{S}) + 2d(M)c(g)\|g - g_n\|.$$

Therefore, as n approaches infinity, $E_{g_n}(\mathscr{S})$ converges to $E_g(\mathscr{S})$. ∎

We are now ready to prove the theorem which we stated previously. That is:

THEOREM 14.

$$\inf_{T \subseteq M} E_T(\mathscr{S}) = \inf_g E_g(\mathscr{S})$$

Proof. We will prove the theorem by proving that, for any g,

$$\inf_{T \subseteq M} E_T(\mathscr{S}) \leq E_g(\mathscr{S}).$$

There is nothing to prove if g has only a finite number of values (by theorem 12).

For a $g(x)$ with an infinite number of values we use the preceding lemma as follows. If $k/n < g(x) \leq (k+1)/n$, we define $g_n(x) = (k+1)/n$. Since we are assuming $\|g\| \leq 1$, $g_n(x)$ has at most $2n+1$ values. We also have $\|g_n\| \leq 1$ and $\|g - g_n\| \leq 1/n$. For each n, we can find a $T_n \subseteq M$ such that $E_{T_n}(\mathscr{S}) \leq E_{g_n}(\mathscr{S})$. Then

$$\inf_T E_T(\mathscr{S}) \leq \inf_{T_n} E_{T_n}(\mathscr{S}) \leq \liminf_{T_n} E_{T_n}(\mathscr{S}) \leq \liminf E_{g_n}(\mathscr{S}) = E_g(\mathscr{S})$$

and the theorem is proven. ∎

Thus in choosing a best 2-dimensional space of approximating functions we need only look at characteristic functions. However, we do not know yet whether a best 2-dimensional space exists. What we would like to know is: does there exist a $T_1 \subseteq M$ such that

$$E_{T_1}(\mathscr{S}) = \inf_{T \subseteq M} E_T(\mathscr{S}) \,?$$

By using theorem 11, the question becomes the following metric space question: Does there exist a $T_1 \subseteq M$ such that

$$\max(d(T_1), (T_1^c)) = \inf_{T \subseteq M} \max(d(T), d(T^c))?$$

This question is answered by the following

THEOREM 15. *Let $T_n \subseteq M$ be such that $\lim_{n \to \infty} \max[d(T_n) d(T_n^c)] = D$. Then there is a $T \subseteq M$ such that $\max[d(T), d(T^c)] \leq D$.*

Proof. Pick any $x_0 \in M$. By possibly interchanging T_n and T_n^c, we can assume $x_0 \in T_n$. Let $f_n(x) = \rho(x, T_n) = \inf_{y \in T_n} \rho(x, y)$. Then $0 \leq f_n(x) \leq \rho(x, x_0) \leq d(M)$ and thus $\{f_n\}$ is uniformly bounded. Now let $x, y \in M$, and $\delta > 0$ be given. There is a $y_1 \in T_n$ such that

$$|f_n(y) - \rho(y, y_1)| < \delta.$$

Then

$$f_n(x) - f_n(y) = \inf_{z \in T_n} \rho(x, z) - \inf_{z \in T_n} \rho(y, z)$$
$$\leq \rho(x, y_1) - \rho(y, y_1) + \delta$$
$$\leq \rho(x, y) + \delta.$$

Similarly, we could get $f_n(y) - f_n(x) \leq \rho(x, y) + \delta$. Since δ is arbitrary, we get $|f_n(x) - f_n(y)| \leq \rho(x, y)$ and thus $\{f_n\}$ is equicontinuous. By the Ascoli-Arzela theorem, $\{f_n\}$ has a uniformly convergent subsequence which we assume is $\{f_n\}$ itself. Let f be the limit. Let $T = f^{-1}(0)$. T is non-empty since $x_0 \in T$. Our theorem will be proven once we show max $(d(T), d(T^c)) \leq D$.

To prove $d(T) \leq D$ we take $x, y \in T$ or x, y such that $f(x) = f(y) = 0$. Given $\delta > 0$ we can find N such that $n \geq N \Rightarrow f_n(x) < \delta, f_n(y) < \delta$ and $d(T_n) \leq D + \delta$. We can find $x_1, y_1 \in T_n$ such that $\rho(x, x_1) < \delta$ and $\rho(y, y_1) < \delta$. Thus

$$\rho(x, y) \leq \rho(x, x_1) + \rho(x_1, y_1) + \rho(y_1, y) < \delta + (D + \delta) + \delta = D + 3\delta.$$

Since δ was arbitrary, $\rho(x, y) \leq D$ and thus $d(T) \leq D$.

To prove $d(T^c) \leq D$ we take $x, y \in T^c$, i.e., $f(x) \neq 0$ and $f(y) \neq 0$. Given $\delta > 0$, we can find an n such that $f_n(x) > 0, f_n(y) > 0$ and $d(T_n^c) < D + \delta$. Therefore, $\rho(x, T_n) > 0$ (or $x \in T_n^c$) and $\rho(y, T_n) > 0$ (or $y \in T_n^c$). Then $\rho(x, y) \leq d(T_n^c) < D + \delta$ and we have $d(T^c) \leq D$. ∎

CHAPTER IX

JACKSON'S THEOREM IN k DIMENSIONS

In this chapter, we turn our attention to the proving of a Jackson theorem for some sets in R^k. We shall also prove that polynomials are an essentially best space by proving a lower bound of the same order of magnitude as exhibited in the Jackson theorem.

The novelty of our approach will arise from the following consideration. We will be considering polynomials of degree n in k variables. Thus we have two parameters, k and n. Strictly speaking, the main consideration in a Jackson theorem is the dependence on n and any dependence on k can be subsumed into a constant term (which, of course, would change with k). If we were then to find a lower bound in terms of k and n, which, when considered as a function of n alone, would have the same order of magnitude as gotten in the Jackson theorem, we will have proven that polynomials are an essentially best approximating space in each of the individual sets under consideration.

However, we will find a function of n and k which (except for a multiplicative constant independent of k and n) will be the same for a lower bound and an upper bound for the degree of approximation by polynomials. Thus we will end up proving that polynomials are a *uniformly* best approximating space.

Notation. (1) R^k will denote real Euclidean k space.

(2) For $x = (x_1, x_2, \ldots, x_k)$ and $y = (y_1, \ldots, y_k)$ in R^k, the inner product of x and y, denoted by $x \cdot y$, will be $\sum_{i=1}^{k} x_i y_i$ and $\|x\| = (x \cdot x)^{1/2}$.

(3) For $k \geq 2$, S_{k-1} will be the unit $k-1$ sphere or
$$S_{k-1} = \{x : x \in R^k, \|x\| = 1\}.$$

(4) B_k will be the unit k ball or $B_k = \{x : x \in R^k, \|x\| \leq 1\}$.

(5) W_k will be the unit k cube or

$$W_k = \{x : x = (x_1, \ldots, x_k) \in R^k \text{ and } |x_i| \leq 1 \text{ for } i = 1, \ldots, k\}.$$

(6) $\mathscr{P}_n^k$ will be the space of polynomials of degree at most n in $x_1, \ldots, x_k$ or $\mathscr{S}_n^k$ will be the space of all linear combinations of $x_1^{\alpha_1} x_2^{\alpha_2} \cdots x_k^{\alpha_k}$ with $\sum_{i=1}^k \alpha_i \leq n$, $\alpha_i \geq 0$.

(7) For $T \subset R^k$, $E_n(\mathscr{S}(T))$ will denote $E_{\mathscr{P}_n^k}(\mathscr{S}(T))$.

(8) For $T \subset R^k$, $C(T)$ will be the space of continuous functions on T.

(9) All functions, constants, etc. will be assumed to be real.

LEMMA 1. *The dimension of $\mathscr{S}_n^k$ is $\binom{n+k}{k}$*

Proof. Since every element of $\mathscr{S}_n^k$ is a linear combination of $x_1^{\alpha_1} \cdots x_k^{\alpha_k}$, we just have to count the number of different ways we can pick out $(\alpha_1, \alpha_2, \ldots, \alpha_k)$ such that $\alpha_i \geq 0$ and $\sum_{i=1}^k \alpha_i \leq n$. For each such choice of $(\alpha_1, \alpha_2, \ldots, \alpha_k)$ we consider the mapping of

$$(\alpha_1, \alpha_2, \ldots \alpha_k) \to (\alpha_1 + 1, (\alpha_1 + 1) + (\alpha_2 + 1), \ldots, \sum_{i=1}^k \alpha_i + k).$$

This is a 1-1 mapping onto the set of k different numbers chosen from 1 to $n+k$ (ordered in increasing order). Since the dimension of the range is $\binom{n+k}{k}$, the dimension of the domain is $\binom{n+k}{k}$. ∎

Thus, the dimension of $\mathscr{S}_n^k$ when restricted to either B_k or W_k is $\binom{n+k}{k}$. However, when we restrict $\mathscr{S}_n^k$ to S_{k-1}, then by the relationship $x_k^2 = 1 - x_1^2 - \cdots - x_{k-1}^2$, we only have to consider those cases for which $\alpha_k = 0$ or 1. In the first case, by our lemma, we have dimension $\binom{n+k-1}{k-1}$ and in the second case we have dimension $\binom{n-1+k-1}{k-1}$ or $\binom{n+k-2}{k-1}$. Altogether, the dimension of $\mathscr{S}_n^k$ restricted to S_{k-1} is $\binom{n+k-1}{k-1} + \binom{n+k-2}{k-1} = \binom{n+k-2}{k-1}(2n+k-1)/n$.

(1) Lower Bounds

Our plan for proving polynomials are essentially best approximating spaces on each of the sets S_{k-1}, B_k, and W_k is as follows: We will first get lower bounds on ε_n for each of the sets (and thus lower bounds for the approximation by any approximating space). Then we will get upper bounds for approximating by polynomials for each of the sets, which will have the same order of magnitude (in terms of n and k) as the previously obtained lower bounds.

To get each of the lower bounds we need to use arguments about the packing of spheres. We will not strive to get the best packing but will be content with the correct order of magnitude.

LEMMA 2. *If $d^k n < 1$, then $\varepsilon_n(B_k) \geq d$.*

Proof. Let $v(r)$ be the volume of a k-ball of radius r. Take any point

$x_1 \in B_k$ and remove from B_k all points within a distance d of x_1. Then choose x_2 in the remaining set and remove from the remaining set all points within a distance d from x_2. If we have chosen n points in this manner and $nv(d) < v(1)$ we can choose another point. However, since $v(d) = d^k v(1)$, we can continue as long as $nd^k < 1$ and thus get $n+1$ points of mutual distance at least d. Thus, $\varepsilon_n(B_k) \geq d$ if $nd^k < 1$.

THEOREM 3. $E_n(\mathscr{S}(B_k)) \geq (1/2e)(k/(n+k))$.

Proof. Since the dimension of $\mathscr{P}_n^k$ restricted to B_k is $\binom{n+k}{n} = n_k$, we know that $E_n(\mathscr{S}(B_k)) \geq (1/2)\varepsilon_{n_k}(B_k)$. However, by lemma 2, as long as $d^k n_k < 1$, $\varepsilon_{n_k}(B_k) \geq d$. Therefore, for any $d < (n_k)^{-1/k}$, $\varepsilon_{n_k}(B_k) \geq d$. Thus $\varepsilon_{n_k}(B_k) \geq (n_k)^{-1/k}$ and we just have to prove that

$$(n_k)^{-1/k} \geq \frac{1}{e} \frac{k}{n+k}.$$

However, $\binom{n+k}{k} < (n+k)^k/k! < (e/k)^k (n+k)^k$ and thus

$$(n_k)^{-1/k} > \frac{1}{e} \frac{k}{n+k}. \quad \blacksquare$$

We next get a lower bound on $E_n(\mathscr{S}(S_{k-1}))$. First, though we need the following

LEMMA 4. *If $d^k n < 1$ then $\varepsilon_n(S_k) \geq d$.*

Proof. By lemma 2, we can find $n+1$ points of mutual distance at least d in B_k. Call these points $x_i = (x_{i1}, x_{i2}, \ldots, x_{ik})$ $i = 1, 2, \ldots, n+1$. Let $z_i = \sqrt{1 - \|x_i\|^2}$ and let $y_i = (x_{i1}, \ldots, x_{ik}, z_i)$. Then the y_i are points of S_k and are of mutual distance at least d. Thus $\varepsilon_n(S_k) \geq d$.

THEOREM 5. $E_n(\mathscr{S}(S_{k-1})) \geq (1/12e)(k/n+k)$.

Proof. Since the dimension of $\mathscr{P}_n^k$ in S_{k-1} is

$$\binom{n+k-2}{k-1} \frac{(2n+k-1)}{n} = n_k^*,$$

we know that $E_n(\mathscr{S}(S_{k-1})) \geq (1/2)\varepsilon_{n_k^*}(S_{k-1})$. By lemma 2, and the argument used in the previous theorem, we know that

$$\varepsilon_{n_k^*}(\mathscr{S}(S_{k-1})) \geq (n_k^*)^{-1/(k-1)}.$$

Thus we just have to show $(n_k^*)^{-1/(k-1)} \geq (1/6e)(k/(n+k))$. However,

$$n_k^* = \binom{n+k-2}{k-1} \frac{2n+k-1}{n}$$

$$\leq \frac{(n+k-2)^{k-1}}{(k-1)!} \frac{2n+k-1}{n}$$

$$\leq \left(\frac{e}{k-1}\right)^{k-1} (n+k-2)^{k-1} \frac{(2n+k-1)}{n}$$

$$\leq \left(\frac{3e(n+k-2)}{k-1}\right)^{k-1}$$

and thus

$$(n_k^*)^{-1/k-1} \geq \frac{k-1}{(3e)(n+k-2)} \geq \frac{1}{6e}\frac{k}{n+k} \blacksquare$$

Finally we will get a lower bound on $E_n(\mathscr{S}(W_k))$. Again we will need a

LEMMA 6. $\varepsilon_n(W_k) \geq (\sqrt{k}/5) n^{-1/k}$.

Proof. We will proceed as in lemma 2. Thus, since the volume of a ball of radius r is

$$\frac{\pi^{k/2} r^k}{\Gamma\left(\frac{k}{2}+1\right)}$$

we certainly can find $n+1$ points of mutual distance at least $(\sqrt{k}/5) n^{-1/k}$, if

$$\frac{n\pi^{k/2}\left(\frac{\sqrt{k}}{5} n^{-1/k}\right)^k}{\Gamma\left(\frac{k}{2}+1\right)} < 2^k$$

However, that is equivalent to

$$\frac{\pi^{k/2}\left(\frac{\sqrt{k}}{10}\right)^k}{\Gamma\left(\frac{k}{2}+1\right)} < 1$$

which follows from a trivial estimate on the gamma function. $\blacksquare$

As before, by using the lemma, we can get a lower bound on $E_n(\mathscr{S}(W_k))$,

THEOREM 7. $E_n(\mathscr{S}(W_k)) \geq (1/10e)(k^{3/2}/(n+k))$.

Proof.

$$E_n(\mathscr{S}(W_k)) \geq \frac{1}{2}\varepsilon_{n_k}(W_k)$$
$$\geq \frac{\sqrt{k}}{10}(n_k)^{-1/k}$$
$$> \frac{\sqrt{k}}{10e}\frac{k}{n+k}$$

by using the estimate from theorem 3. ∎

(2) Upper Bounds

With this theorem, we have completed the derivation of the lower bounds. We now proceed to get the upper bounds for each of the three cases. However, it will turn out that once we have the upper bound for $E_n(\mathscr{S}(S_{k-1}))$ we shall easily derive the other upper bounds from that one. Accordingly, we shall state the result for $E_n(\mathscr{S}(S_{k-1}))$, use it to get the other upper bounds and then return and prove the upper bound for $E_n(\mathscr{S}(S_{k-1}))$.

THEOREM 8. $E_n(\mathscr{S}(S_{k-1})) \leq 44(k/(k+n))$.

Proof. Later.

In terms of this theorem and by using theorem 8.7, we can get an estimate on $E_{\mathscr{P}_n^k}(f)$ for any $f \in C(S_{k-1})$.

Corollary. $E_{\mathscr{P}_n^k}(f) \leq 5\omega(f, 44(k/(k+n)))$ for $f \in C(S_{k-1})$.

We shall now use theorem 8 to prove the following

THEOREM 9. $E_n(\mathscr{S}(B_k)) \leq 88(k/(n+k))$.

Proof. Let $f \in \mathscr{S}(B_k)$ and let $g(x_1, \ldots, x_{k+1}) = f(x_1, \ldots, x_k)$ where

$$x_{k+1}^2 = 1 - \sum_{i=1}^{k} x_i^2$$

We then have

$$|g(x_1, \ldots, x_{k+1}) - g(y_1, \ldots, y_{k+1})| = |f(x_1, \ldots x_k) - f(y_1, \ldots, y_k)|$$
$$\leq \sqrt{\sum_{i=1}^{k}(x_i - y_i)^2}$$
$$\leq \sqrt{\sum_{i=1}^{k+1}(x_i - y_i)^2}$$

and thus $g \in \mathscr{S}(S_k)$. By theorem 8, we know there exists a $P \in \mathscr{P}_n^{k+1}$ such that $|g(x) - P(x)| \leq 44((k+1)/(k+1+n))$ for $x \in S_k$. Since

g is even in the variable x_{k+1}, this inequality remains true even if we replace $P(x)$ by

$$Q(x) = \frac{P(x_1, x_2, \ldots, x_{k+1}) + P(x_1, \ldots, -x_{k+1})}{2}$$

Since $Q(x)$ is now even in the variable x_{k+1}, we can replace x_{k+1}^2 by $1 - \sum_{i=1}^{k} x_i^2$ and we call the new polynomial $R(x)$. Then $R(x) \in \mathscr{P}_n^k$ and since, for $x \in B_k$,

$$|f(x) - R(x)| \leq 44\left(\frac{k+1}{k+1+n}\right) \leq 88\left(\frac{k}{k+n}\right)$$

we have proven the theorem. ∎

As before, we have a

Corollary. $E_{\mathscr{P}_n^k}(f) \leq 5\omega(f, 88(k/(n+k)))$ for $f \in C(B_k)$.

Now that we have established an upper bound for $E_n(\mathscr{S}(B_k))$ (which is based on the one for $E_n(\mathscr{S}(S_{k+1}))$) we can get one for $E_n(\mathscr{S}(W_k))$.

THEOREM 10. $E_n(\mathscr{S}(W_k)) \leq 88(k/(n+k))$.

Proof. Let D_k be the ball $\{x : x \in R^k, \|x\| \leq \sqrt{k}\}$. Then $W_k \subset D_k$. Let $f \in \mathscr{S}(W_k)$. Then we know, that f can be extended to a function (which we still call f) which is in $\mathscr{S}(D_k)$. Let $g(x) = f(\sqrt{k}\,x)$ for $x \in B_k$. Then $(1/\sqrt{k})g(x) \in \mathscr{S}(B_k)$. By theorem 9, we know there exists a $P \in \mathscr{P}_n^k$ such that $|P(x) - (1/\sqrt{k})g(x)| \leq 88(k/(k+n))$ for $x \in B_k$. Letting $Q(x) = \sqrt{k}\,P(x/\sqrt{k})$ we have that

$$\left|Q(x) - g\left(\frac{x}{\sqrt{k}}\right)\right| \leq 88\frac{k^{3/2}}{k+n}$$

for $x \in D_k$ or

$$\left|Q(x) - f(x)\right| \leq 88\frac{k^{3/2}}{k+n}$$

for $x \in D_k$ and certainly for $x \in W_k \subset D_k$. ∎

Once again, we have a

Corollary. For $f \in C(W_k)$, $E_{\mathscr{P}_n^k}(f) \leq 5\omega(f, 88(k^{3/2}/(k+n)))$

(3) Proof of Theorem 8

We now proceed to prove theorem 8. First, though, we will need four lemmas.

LEMMA 11. Let (M, ρ), a compact metric space, have the property that whenever $\rho(x, y) = a + b$, where $a > 0$ and $b > 0$, there exists a

$z \in M$ such that $\rho(x, z) = a$ and $\rho(z, y) = b$. Let L be a positive linear operator taking $C(M)$ into $C(M)$ such that $(L(1))(x) \equiv 1$. Then for any $f \in C(M)$ and any $t > 0$,

$$|(Lf)(y) - f(y)| \leq \omega(f, t)\left\{1 + \frac{((L\rho_y^2)(y))^{1/2}}{t}\right\}$$

where ρ_y is the function $\rho_y(x) = \rho(y, x)$.

Proof. Given any x, y in M, we have that $\rho(x, y) = t + (\rho(x, y) - t)$. Thus, there exists (if $t < \rho(x, y)$) a $z_1 \in M$ such that $\rho(x, z_1) = t$ and $\rho(z_1, y) = \rho(x, y) - t = t + (\rho(x, y) - 2t)$. Then (if $2t < \rho(x, y)$), we find a $z_2 \in M$ such that $\rho(z_1, z_2) = t$ and $\rho(z_2, y) = \rho(x, y) - 2t$. We continue in this manner and get z_j such that $\rho(z_j, z_{j-1}) = t$ and $\rho(z_j, y) = \rho(x, y) - jt$ as long as $\rho(x, y) - jt > 0$. We end up with $[(\rho(x, y)/t)] = m$ points z_j. Then

$$|f(x) - f(y)| \leq |f(x) - f(z_1)|$$
$$+ |f(z_1) - f(z_2)| + \cdots + |f(z_m) - f(y)|$$
$$\leq (m + 1)\omega(f, t)$$

Thus

$$|f(x) - f(y)| \leq \left(1 + \frac{\rho_y(x)}{t}\right)\omega(f, t).$$

Applying L to this inequality (and keeping y fixed) we get

$$|(Lf)(x) - f(y)| \leq \omega(f, t)\left\{1 + \frac{(L\rho_y)(x)}{t}\right\}$$
$$\leq \omega(f, t)\left\{1 + \frac{(L\rho_y^2)(x)^{1/2}}{t}\right\}$$

(by the Schwarz inequality for positive linear operators). If we now take y to be x, we get the desired inequality. ∎

The next lemma is very similar to the methods of theorem 4.1.

LEMMA 12. *Let k be a fixed even integer and let $\{P_i(t)\}$ be the orthogonal polynomials (over $[-1, 1]$) corresponding to the weight function $(1 - t^2)^{(k-3)/2}$. Then, as $P(t)$ ranges over all polynomials of degree at most $2m$ which are non-negative for $t \in [-1, 1]$ the maximum of*

$$\frac{\int_{-1}^{1} tP(t)(1 - t^2)^{(k-3)/2} dt}{\int_{-1}^{1} P(t)(1 - t^2)^{(k-3)/2} dt}$$

is equal to λ_{m+1} which is the largest root of $P_{m+1}(t)$.

Proof. By methods very similar to those of Chapter IV, one can easily derive the quadrature formula

$$\int_{-1}^{1} P(t)(1-t^2)^{(k-3)/2} dt = \sum_{i=1}^{n} c_i P(\lambda_i) \quad \text{for all} \quad P \in \mathscr{S}_{2n-1}$$

where $\lambda_1, \lambda_2, \ldots, \lambda_n$ are the zeroes of P_n (in order) and $c_i > 0$. Since, in our case, both $tP(t)$ and $P(t)$ are in $\mathscr{S}_{2m+1}$ we get that

$$\frac{\int_{-1}^{1} tP(t)(1-t^2)^{(k-3)/2} dt}{\int_{-1}^{1} P(t)(1-t^2)^{(k-3)/2}} = \frac{\sum_{i=1}^{m+1} c_i \lambda_i P(\lambda_i)}{\sum_{i=1}^{m+1} c_i P(\lambda_i)}.$$

Since $P(\lambda_i) \geq 0$, the numerator is $\leq \lambda_{m+1} \sum_{i=1}^{m+1} c_i P(\lambda_i)$ which gives us an upper bound of λ_{m+1}. Furthermore, by taking

$$P(t) = \left(\frac{P_{m+1}(t)}{t - \lambda_{m+1}}\right)^2$$

we can actually get λ_{m+1}. ∎

The next lemma gives us an estimate (from below) on λ_{m+1}.

LEMMA 13. Let λ_{m+1} be as in the previous lemma. Then

$$\lambda_{m+1} \geq 1 - 2\pi^2 \left(\frac{k}{2m+k}\right)^2$$

Proof. We set $k = 2r$ and then $(k-3)/2 = (r-1) - 1/2$. We set $s = m + r$ and let $t_1 \geq t_2 \geq \cdots \geq t_{k-1}$ be the $k-1$ largest roots of $T_s(x)$, the Tchebychev polynomial of degree s. We now consider

$$\int_{-1}^{1} \frac{(T_s(t))^2 (1-t^2)^{(k-3)/2} dt}{(t_i - t) \cdots (t_{k-1} - t)} = \int_{-1}^{1} T_s(t) \left[\frac{T_s(t)(1-t^2)^{r-1}}{(t_1 - t) \cdots (t_{k-1} - t)}\right] \frac{dt}{\sqrt{1-t^2}}$$

Since, in the square brackets, we have a polynomial of degree at most $(s + 2r - 2) - (k - 1) = s - 1$ which we are integrating against $T_s(t)/(\sqrt{1-t^2})$, we get that the integral is 0 (we leave as a trivial exercise the fact that the Tchebychev polynomials are orthogonal polynomials with respect to the weight function $1/(\sqrt{1-t^2})$). On the other hand, since

$$Q(t) = \frac{T_s(t)^2}{(t_1 - t) \cdots (t_{k-1} - t)}$$

is a polynomial of degree at most $2s - (k-1) = 2m + 1$, we may evaluate the integral by means of the quadrature formula using the zeroes of P_{m+1}. We thus have $\sum_{i=1}^{m+1} c_i Q(\lambda_i) = 0$. We now claim that $t_{k-1} \leq \lambda_{m+1}$. If not, then all the λ_i would be in $[-1, t_{k-1})$ and since $Q(t) \geq 0$ for $t \in [-1, t_{k-1}]$ we would get that $Q(\lambda_i) = 0$ (recall that $c_i > 0$) and, moreover, by the non-negativity of $Q(t)$ there, they would all be double roots. Thus we would get $2(m+1)$ roots of a non-identically zero polynomial of degree at most $2m + 1$ which is impossible. Thus

$$\lambda_{m+1} \geq t_{k-1} = \cos\left(\frac{\pi(2k-3)}{2m+k}\right) \geq 1 - \frac{1}{2}\left(\frac{\pi(2k-3)}{2m+k}\right)^2$$
$$> 1 - 2\pi^2\left(\frac{k}{2m+k}\right)^2$$

and the lemma is proven. ∎

We just have one more lemma before we prove theorem 8. It is a standard result on integration in S_{k-1} and the proof can be found in ([8] p. 247).

LEMMA 14. Let $f(t)$ be integrable on $[-1, 1]$ and let $d\sigma$ denote surface measure. Then

$$\int_{S_{k-1}} f(x \cdot y) d\sigma(x) = C_k \int_{-1}^{1} f(t)(1 - t^2)^{(k-3)/2} dt$$

where C_k is a constant depending only on k.

We are now ready to prove theorem 8.

Proof of theorem 8. What we will actually prove is that for n even and k even, $E_n(\mathscr{S}(S_{k-1})) \leq 11(k/(n+k))$. Then, if n is odd, we would have

$$E_n(\mathscr{S}(S_{k-1})) \leq E_{n-1}(\mathscr{S}(S_{k-1})) \leq 11\frac{k}{k+(n-1)} \leq 22\frac{k}{k+n}$$

and end up with $E(\mathscr{S}(S_{k-1})) \leq 22(k/(k+n))$ for k even and any n. Then, for k odd we would have the following consideration. We can consider S_{k-1} imbedded in S_k since, if $x = (x_1, \ldots, x_k) \in S_{k-1}$ then $(x_1, \ldots, x_k, 0) \in S_k$. Thus, any $f \in \mathscr{S}(S_{k-1})$ can be extended to an $f \in \mathscr{S}(S_k)$. Then, since $k + 1$ is now even, we know there exists a $P \in \mathscr{T}_n^{k+1}$ such that

$$|f(x_1, \ldots, x_k, x_{k+1}) - P(x_1, \ldots, x_k, x_{k+1})| \leq 22\frac{k+1}{k+1+n}$$

for all $(x_1, \ldots, x_k, x_{k+1})$ in S_k. In particular, we would have that

$$|f(x_1,\ldots,x_k,0) - P(x_1,\ldots,x_k,0)| \leq 22\,\frac{k+1}{k+1+n}$$

for $(x_1,\ldots,x_k,0)$ in S_k or

$$|f(x_1,\ldots,x_k) - P(x_1,\ldots,x_k,0)| \leq 22\,\frac{k+1}{k+1+n}$$

for $(x_1,\ldots,x_k)$ in S_{k-1}. Since $P(x_1,\ldots,x_k,0) \in \mathscr{I}_n^k$ and f is an arbitrary element $\varepsilon \mathscr{S}(S_{k-1})$, we have that

$$E_n(\mathscr{S}(S_{k-1})) \leq 22\,\frac{k+1}{k+1+n} \leq 44\,\frac{k}{n+k}.$$

We thus only have to prove that $E_n(\mathscr{S}(S_{k-1})) \leq 11(k/(k+n))$ for n and k even. In order to satisfy the hypothesis of lemma 11, we consider S_{k-1} as a metric space with metric $\rho(x,y)$ measured along the minor arc of the circle in which the 2-plane determined by x, y and 0 cuts S_{k-1}. Then $\|x-y\| \leq \rho(x,y) \leq \pi/2\,\|x-y\|$ so that the modulus of continuity factor in lemma 11 is unaffected by the change in metric. We let $n = 2m$, $\{P_i(t)\}$ be the orthogonal polynomials determined by the weight function $(1-t^2)^{(k-3)/2}$, λ_{m+1} be the largest root of $P_{m+1}(t)$ and set

$$P(t) = \left(\frac{P_{m+1}(t)}{t - \lambda_{m+1}}\right)^2.$$

We define the positive, linear operator $L: C(S_{k-1}) \to C(S_{k-1})$ by

$$(Lf)(x) = \frac{\int_{S_{k-1}} P(x \cdot z) f(z) d\sigma(z)}{\int_{S_{k-1}} P(x \cdot z) d\sigma(z)}$$

We note that the denominator is independent of x, $(L1)(x) \equiv 1$ and $(Lf)(x) \in \mathscr{I}_n^k$. By lemma 11 (with $t = k/(n+k)$), we see that if

$$(L\rho_y^2)(y) \leq 100\left(\frac{k}{n+k}\right)^2$$

we will have proven the theorem. However,

$$(\rho_y^2)(x) = \rho^2(x,y) \leq \frac{\pi^2}{4}\|x-y\|^2 \leq 5(1-(x \cdot y)).$$

Thus,

$$(L\rho_y^2)(y) \leq 5(1 - (L(x \cdot y))(y)) = 5\left\{1 - \frac{\int_{S_{k-1}} P(y \cdot z)(y \cdot z) d\sigma(x)}{\int_{S_{k-1}} P(y \cdot z) d\sigma(z)}\right\}$$

$$= 5\left\{1 - \frac{\int_{-1}^{1} tP(t)(1-t^2)^{(k-3)/2} dt}{\int_{-1}^{1} P(t)(1-t^2)^{(k-3)/2} dt}\right\}$$

$$= 5\{1 - \lambda_{m+1}\}$$

$$\leq 10\pi^2 \left(\frac{k}{k+n}\right)^2 < 100\left(\frac{k}{k+n}\right)^2 \blacksquare$$

CHAPTER X

COMPLETENESS

In the last few chapters, we've considered one kind of question; that is the question "How well can we approximate?" That, of course, is the kind of question that the Jackson theorem answered.

In this chapter, we take a temporary detour from that kind of question and instead concern ourselves with the general question of "Can we approximate arbitrarily closely?" That, of course, is the kind of question answered by the Weierstrass theorem. Aside from introducing us to new problems with interesting methods of solution, it also introduces us to the Müntz theorem which, in a later chapter, we consider in the form of a Müntz-Jackson theorem.

(1) Completeness

We first have the following

Definition. A set of vectors, $\{V_n\}$, in a Banach space B, is said to be complete in B, if every vector $X \in B$ can be approximated to any degree of accuracy by finite linear combinations of the V_n.

The next theorem gives an alternative definition of completeness.

THEOREM 1. $\{V_n\}$ is complete in B iff there does not exist a non-trivial bounded linear functional L on B such that $L(V_n) = 0$ for all n.

Proof. Take any $X \in B$ and assume $\{V_n\}$ is complete in B. Then, for given $\varepsilon > 0$, we can find $a_1, \ldots, a_k$ such that

$$\|X - (a_1 V_1 + \cdots + a_k V_k)\| < \varepsilon.$$

Take any bounded linear functional L on B such that $L(V_n) = 0$ for all n. Then $|L(X)| = |L(X - (a_1 V_1 + \cdots + a_k V_k))| \leq \|L\| \varepsilon$. Since ε was arbitrary, we have that $L(X) = 0$. Since X was an arbitrary element of B, we have that $L(X) = 0$ for all $X \in B$. Thus, there is no non-trivial bounded linear functional L on B such that $L(V_n) = 0$ for all n.

Now assume $\{V_n\}$ is not complete in B and we will prove the existence of a non-trivial L such that $L(V_n) = 0$ for all n.

We know there is an $X_0 \in B$ such that

$$\inf_{Y_k} \|X_0 - Y_k\| = d > 0$$

where the Y_k are chosen from among the finite linear combinations of $\{V_n\}$. Let B_1 be the subspace of B of all vectors of the form

$$\alpha X_0 + a_1 V_1 + \cdots + a_n V_n.$$

That is, B_1 is the space of all finite linear combinations of X_0 and V_n. Let L be defined on B_1 by

$$L(\alpha X_0 + a_1 V_1 + \cdots + a_n V_n) = \alpha.$$

(This is a definition since in representing a vector from B_1 in the form $\alpha X_0 + a_1 V_1 + \cdots + a_n V_n$ the α is unique.) We know that

$$\|\alpha X_0 + a_1 V_1 + \cdots + a_n V_n\| \geq |\alpha| d = |L(\alpha X_0 + \cdots + a_n V_n)| d$$

and thus

$$\|L\| \leq \frac{1}{d}.$$

Let $\{Y_k\}$ be such that $\|X_0 - Y_k\|$ converges to d where Y_k is a finite linear combination of $\{V_n\}$. Then

$$1 = |L(X_0 - Y_k)| \leq \|L\| \, \|X_0 - Y_k\|.$$

Thus $\|L\| \geq 1/\|X_0 - Y_k\|$ which converges to $1/d$. Therefore $\|L\| \geq 1/d$ and consequently $\|L\| = 1/d$.

By the Hahn-Banach theorem, L can be extended to be a bounded linear functional on B such that $L(V_n) = 0$ and $\|L\| = 1/d$. We have thus proven the theorem. ∎

The previous theorem is usually used together with the well known

RIESZ REPRESENTATION THEOREM:
(1) L is a bounded linear functional on $C[a, b]$ iff there is a finite measure μ on $[a, b]$ such that

$$L(f) = \int_a^b f(x) d\mu(x)$$

for all $f \in C[a, b]$.

(2) For $1 < P < \infty$, L is a bounded linear functional on $L^P[a, b]$ iff there is a $g \in L^q[a, b]$ where $1/P + 1/q = 1$ and

$$L(f) = \int_a^b f(x)g(x)dx$$

for all $f \in L^P[a, b]$.

(3) L is a bounded linear function on $L^1[a, b]$ iff there is a $g \in L^\infty[a, b]$ such that

$$L(f) = \int_a^b f(x)g(x)d(x)$$

for all $f \in L^1[a, b]$.

The following theorem is easily proved using Morera's theorem.

THEOREM 2. Let $f(z, x)$ be analytic in a region R for all $x \in [a, b]$ and let μ be a finite measure on $[a, b]$. Then $\int_a^b f(z, x)d\mu(x)$ is analytic in R.

EXAMPLE. a) $\int_a^b e^{izx} d\mu(x)$ is entire when $\mu(x)$ is a finite measure on $[a, b]$.

As an example of a theorem proved by using some of these theorems, we have:

THEOREM 3. $\{x^n\}_{n=0}^\infty$ is complete in $L^2[0, 1]$.

Proof. We assume L, a bounded linear functional on $L^2[0, 1]$, has the property that $L(x^n) = 0$, $n = 0, 1, \ldots$. By the Riesz representation theorem, we know that $L(f) = \int_0^1 f(x)g(x)dx$ for some fixed $g \in L^2[0, 1]$ and for all $f \in L^2[0, 1]$. Thus

$$\int_0^1 x^n g(x)\, dx = 0, \quad n = 0, 1, 2 \ldots$$

Therefore,

$$0 = \sum_{n=0}^\infty \int_0^1 \frac{(2\pi i k)^n x^n}{n!} g(x)dx$$

$$= \int_0^1 \left(\sum_{n=0}^\infty \frac{(2\pi i k)^n}{n!} x^n \right) g(x) dx \text{ (by dominated convergence)}$$

$$= \int_0^1 e^{2\pi i k x} g(x)\, dx.$$

Since, all the Fourier coefficients of $g(x)$ are 0, we must have that $g(x) = 0$ a.e. Thus $L(f) = \int_0^1 f(x)g(x)\, dx = 0$ for all $f \in L^2[0, 1]$ and we have completeness. ∎

(2) Müntz's Theorem

Now let λ_i be real numbers ≥ 1. When is $\{x^{\lambda_n}\}_{n=1}^{\infty}$ complete in $L^2[0,1]$? We know that this is equivalent to "when does $f \in L^2[0,1]$ and

$$\int_0^1 x^{\lambda_n} f(x)\,dx = 0$$

imply $f(x) = 0$ a.e.?" Assume we have such an f and let

$$\varphi(z) = \int_0^1 x^z f(x)\,dx.$$

Then $\varphi(z)$ is analytic in $\operatorname{Re} z > 1/2$ and $\varphi(\lambda_n) = 0$. Let $\operatorname{Re} z \geq 0$. Then

$$|\varphi(z)| = \left|\int_0^1 x^z f(x)\,dx\right| \leq \int_0^1 |x^z f(x)|\,dx$$

$$\leq \left(\int_0^1 |x^z|^2\,dx\right)^{1/2}\left(\int_0^1 |f(x)|^2\,dx\right)^{1/2}$$

$$= \left(\int_0^1 x^{2\operatorname{Re} z}\,dx\right)^{1/2}\left(\int_0^1 |f(x)|^2\,dx\right)^{1/2}$$

$$= \left(\frac{1}{2\operatorname{Re} z + 1}\right)^{1/2}\left(\int_0^1 |f(x)|^2\,dx\right)^{1/2}$$

$$\leq \left(\int_0^1 |f(x)|^2\,dx\right)^{1/2}$$

Thus $\varphi(z)$ is analytic and uniformly bounded in $\operatorname{Re} z \geq 0$ and vanishes at $\{\lambda_n\}$.

LEMMA 4. *Let $h(z)$ be analytic and bounded in $\operatorname{Re} z > 0$ and continuous on $\operatorname{Re} z \geq 0$. Then $\sup_z |h(z)| = \sup_Y |h(iY)|$.*

Proof. Let $\varepsilon > 0$ and consider $h(z)/(1 + \varepsilon z)$.

It is analytic in the right half plane, continuous on the boundary including infinity, since, as $|z|$ approaches infinity, $|h(z)/(1 + \varepsilon z)|$ approaches 0. Since it is continuous on a compact set, its maximum must be achieved on the boundary. Thus, for $\operatorname{Re} z \geq 0$,

$$\left|\frac{h(z)}{1 + \varepsilon z}\right| \leq \max_Y \left|\frac{h(iY)}{1 + \varepsilon i Y}\right| \leq \sup_Y |h(iY)|$$

If we let ε go to zero, we get

$$|h(z)| \leq \sup_Y |h(iY)|$$

It is now obvious that

$$\sup_{z}|h(z)| = \sup_{Y}|h(iY)| \quad \blacksquare$$

We return now to our function $\varphi(z)$ which is analytic and uniformly bounded in $Re\,z \geq 0$ and vanishes at $\{\lambda_n\}$. Then

$$\frac{\varphi(z)}{\dfrac{z-\lambda_1}{z+\lambda_1}}$$

is analytic in $Re\,z \geq 0$ and is bounded there, since in a closed neighborhood of λ_1 it is a continuous function defined on a compact set and outside that neighborhood, the denominator is bounded away from zero. Thus for $Re\,z \geq 0$

$$\left|\frac{\varphi(z)}{\dfrac{z-\lambda_1}{z+\lambda_1}}\right| \leq \sup_{Y}\left|\frac{\varphi(iY)}{\dfrac{iY-\lambda_1}{iY+\lambda_1}}\right| = \sup_{Y}|\varphi(iY)| = M$$

since

$$\left|\frac{iY-\lambda_1}{iY+\lambda_1}\right| = 1.$$

Similarly

$$\left|\frac{\varphi(z)}{\prod_{i=1}^{n}\dfrac{z-\lambda_i}{z-\lambda_i}}\right| \leq M$$

for any n. For fixed z with $Re\,z \geq 0$, consider

$$\left|\prod_{i=1}^{n}\frac{z-\lambda_i}{z+\lambda_i}\right| = \prod_{i=1}^{n}\left|\frac{z-\lambda_i}{z+\lambda_i}\right|$$

As n increases it is a monotonic decreasing sequence. Therefore,

$$\prod_{i=1}^{n}\left|\frac{z-\lambda_i}{z+\lambda_i}\right|$$

either converges, or diverges to 0. If $\varphi(z) \neq 0$, it can't diverge to zero since

$$\left|\frac{\varphi(z)}{\prod_{i=1}^{n}\dfrac{z-\lambda_i}{z+\lambda_i}}\right| \leq M$$

for all n.

In particular, if $\varphi(z) \not\equiv 0$, then there is a $C \in [1/2, 1)$ such that $\varphi(C) \neq 0$. Then
$$\prod_{i=1}^{\infty} \left| \frac{C - \lambda_i}{C + \lambda_i} \right|$$
converges. However, since $\lambda_i > C$,
$$\prod_{i=1}^{\infty} \left| \frac{C - \lambda_i}{C + \lambda_i} \right| = \prod_{i=1}^{\infty} \left(\frac{\lambda_i - C}{\lambda_i + C} \right) = \prod_{i=1}^{\infty} \left(1 - \frac{2C}{\lambda_i + C} \right)$$
which converges iff
$$\sum_{i=1}^{\infty} \frac{1}{\lambda_i}$$
converges. Thus if $\lambda_i \geq 1$ and
$$\sum_{i=1}^{\infty} \frac{1}{\lambda_i} = \infty$$
then $\varphi(z) \equiv 0$ on $[1/2, 1)$ which implies that $\varphi(z) \equiv 0$ or $\int_0^1 x^z f(x) dx \equiv 0$. In particular, let $z = 0, 1, 2, \ldots$. Then $\int_0^1 x^n f(x) dx = 0$ which implies $f(x) = 0$ a.e.

That of course implies that $\{x^{\lambda_i}\}_{i=1}^{\infty}$ is complete in $L^2[0, 1]$.

THEOREM 5. If $-(1/2) < c < \lambda_i$ and $\sum_{\lambda_i \neq 0} |1/\lambda_i| = \infty$, then $\{x^{\lambda_i}\}_{i=1}^{\infty}$ is complete in $L^2[0, 1]$.

Proof. We have to consider two cases:

(a) If the $\{\lambda_i\}$ have no finite limit point, we have only a finite number of $\lambda_i < 1$ and $\sum_{\lambda_i \geq 1} 1/\lambda_i$ diverges. By what we have just proven, $\{x^{\lambda_i}\}_{\lambda_i \geq 1}$ is complete in $L^2[0, 1]$ and certainly $\{x^{\lambda_i}\}_{i=1}^{\infty}$ is complete in $L^2[0, 1]$.

(b) If the $\{\lambda_i\}$ have a finite limit point a, then $a > -(1/2)$. Again, let $\varphi(z) = \int_0^1 x^z f(x) dx$ where $\varphi(\lambda_i) = 0$. $\varphi(z)$ is an analytic function in $\operatorname{Re} z > -(1/2)$ and its zeros have a limit point in its region of analyticity. Thus $\varphi(z) \equiv 0$ and we have seen that is enough to guarantee that $f(x) = 0$ a.e. ∎

Now assume $2 \leq \lambda_n$ and $\sum_{n=1}^{\infty} 1/\lambda_n < \infty$. The condition on these λ_n, that $\sum 1/\lambda_n < \infty$ is necessary and sufficient for
$$\prod_{n=1}^{\infty} \left(\frac{z - \lambda_n}{z + \lambda_n} \right)$$
to converge uniformly on compact subsets of the right half plane. If we set
$$B(z) = \prod_{n=1}^{\infty} \left(\frac{z - \lambda_n}{z + \lambda_n} \right)$$

then $B(z)$ is analytic in $Re z > 0$ and is uniformly bounded there by 1 (since $|(z - \lambda_n)/(z + \bar\lambda_n)| \leq 1$).

Let $\varphi(z) = [1/(z + 1)] \cdot B(z)$. Then $\varphi(z)$ is analytic in the right half plane. Let $x > 0$. Then

$$\int_{-\infty}^{\infty} |\varphi(x + iy)|^2 dy \leq \int_{-\infty}^{\infty} \frac{1}{(x + 1)^2 + y^2} dy$$

$$= \frac{1}{x + 1} \int_{-\infty}^{\infty} \frac{dy}{1 + y^2}$$

$$= \frac{\pi}{x + 1} < \pi.$$

Thus $\int_{-\infty}^{\infty} |\varphi(x + iy)|^2 dy$ is uniformly bounded for $x > 0$. We now rely on the famous

PALEY WIENER THEOREM. A function $f(z)$ is analytic in the right half plane and $\int_{-\infty}^{\infty} |f(x + iy)|^2 dy$ is bounded for all $x > 0$ iff $f(z) = \int_0^{\infty} g(t) e^{-zt} dy$ where $g \in L^2[0, \infty]$.

Using this theorem we have that $\varphi(z) = \int_0^{\infty} g(t) e^{-zt} dt$. If we set $e^{-t} = x$ we get that

$$\varphi(z) = \int_0^1 g(-\log x) x^z \frac{dx}{x}.$$

Call $g(-\log x)/x = h(x)$. Then $h \in L^2[0, 1]$ and $\varphi(z) = \int_0^1 h(x) x^z dx$. Since $\varphi(\lambda_n) = 0$, we have $\int_0^1 h(x) x^{\lambda_n} dx = 0$. Thus $\{x^{\lambda_n}\}$ is incomplete in $L^2[0, 1]$ if $2 \leq \lambda_n$ and $\sum 1/\lambda_n < \infty$.

Now take a sequence $\{\lambda_n\}$ such that $-1/2 < \lambda_n$ and $\sum_{\lambda_n \neq 0} 1/|\lambda_n| < \infty$. Let $\alpha_n = \lambda_n + 3$. Then $\alpha_n > 2$ and $\sum 1/\alpha_n < \infty$. Therefore there is a $g \in L^2[0, 1]$ such that $\int_0^1 g(x) x^{\alpha_n} dx = 0$ for all n. Let $h(x) = g(x) x^3$. Then $h \in L^2[0, 1]$ and $\int_0^1 h(x) x^{\lambda_n} dx = 0$ for all n. ■

We thus have proven the

MÜNTZ THEOREM. Let $\{\lambda_n\}$ be a sequence of real numbers such that $\lambda_n > c > -1/2$. Then $\{x^{\lambda_n}\}$ is complete in $L^2[0, 1]$ iff $\sum_{\lambda_n \neq 0} 1/|\lambda_n| = \infty$.

Corresponding to this theorem about $L^2[0, 1]$ completeness we have the following:

THEOREM 6. Let $1/2 < \lambda_n$. Then $\{1, x^{\lambda_n}\}$ is complete in $C[0, 1]$ iff $\sum 1/\lambda_n = \infty$.

Proof. Assume $\sum 1/\lambda_n = \infty$ and $\int_0^1 x^{\lambda_n} d\mu(x) = \int_0^1 d\mu = 0$. We consider two cases.

(a) There are an infinite number of $\lambda_n \in [1/2, 1]$. Let

$$f(z) = \int_0^1 x^z d\mu(x).$$

Then $f(z)$ is analytic in $\operatorname{Re} z > 0$ and its zeros have a limit point in its domain of analyticity. Therefore $f(z) \equiv 0$. Thus $\int_0^1 x^n d\mu(x) = 0$, $n = 0, 1, 2, \ldots$ and $d\mu \equiv 0$.

(b) There are only a finite number of $\lambda_n \in [1/2, 1]$. Then,

$$\sum_{\lambda_n > 1} \frac{1}{\lambda_n} = \infty$$

and

$$\sum_{\lambda_n > 1} \frac{1}{\lambda_n - 1} \geq \sum_{\lambda_n > 1} \frac{1}{\lambda_n} = \infty.$$

Therefore $\{x^{\lambda_n - 1}\}$ is complete in $L^2[0, 1]$.

Since we know that $\int_0^1 x^{\lambda_n} d\mu(x) = \int_0^1 d\mu(x) = 0$ we can assume that $\mu(0) = \mu(1) = 0$ and that $\mu(x) = (\mu(x^+) + \mu(x^-))/2$.

By integration by parts, we get $\int_0^1 x^{\lambda_n - 1} \mu(x) dx = 0$. $\mu(x)$, a function of bounded variation is certainly in $L^2[0, 1]$. Since $\{x^{\lambda_n - 1}\}$ is complete in $L^2[0, 1]$, we get that $\mu(x) = 0$ a.e. and by our normalization, $\mu(x) \equiv 0$.

Now assume that $\sum (1/\lambda_n) < \infty$. Then $\{x^{\lambda_n}\}$ is incomplete in $L^2[0, 1]$ Assume that $\{1, x^{\lambda_n}\}$ is complete in $C[0, 1]$. Then

$$\left\| x^k - a_0 - \sum_{n=1}^N a_n x^{\lambda_n} \right\| < \frac{\varepsilon}{2}$$

for a given $\varepsilon > 0$.

Evaluating this at $x = 0$ we get $|a_0| < \varepsilon/2$. Thus

$$\left\| x^k - \sum_{n=1}^N a_n x^{\lambda_n} \right\| < \varepsilon$$

Since

$$\left\| x^k - \sum_{n=1}^N a_n x^{\lambda_n} \right\|_{L^2} \leq \left\| x^k - \sum_{n=1}^N a_n x^{\lambda_n} \right\| < \varepsilon$$

we see that we can approximate x^k in $L^2[0, 1]$ for any $k \geq 1$. Since $\{x^n\}_{n=1}^\infty$ is complete in $L^2[0, 1]$ we get that $\{x^{\lambda_n}\}_{n=1}^\infty$ is complete in $L^2[0, 1]$ which is a contradiction. ∎

(3) More Completeness Questions

Actually we could have proven the same result for $\lambda_n \geq C > 0$ and not necessarily $C = 1/2$. What happens, however, if λ_n are all in $[0, 1]$ and converge to 0? This is partly answered by the following

THEOREM 7. If $0 < \lambda_n \leq 1$, then $\{1, x^{\lambda_n}\}_{n=1}^{\infty}$ is complete in $C[0, 1]$ if $\sum \lambda_n = \infty$.

Proof. Assume $\int_0^1 x^{\lambda_n} d\mu(x) = \int_0^1 d\mu(x) = 0$. Let $f(z) = \int_0^1 x^{1/(z+1)} d\mu(x)$. Then $f(z)$ is analytic in $\operatorname{Re} z > -1$ and in particular in $\operatorname{Re} z \geq 0$. Moreover,

$$|f(z)| \leq \|x^{1/(z+1)}\| V$$
$$= V$$

(where V is the total variation of μ). Let $\alpha_n = 1/\lambda_n$. Then $\sum 1/\alpha_n = \infty$. Therefore $\sum_{\alpha_n \neq 1} 1/(\alpha_n - 1) = \infty$.

Combining this with $f(z)$, a bounded analytic function in $\operatorname{Re} z \geq 0$, vanishing at $\alpha_n - 1$, we get that $f(z) \equiv 0$. Setting $z = (1/n) - 1$ we get that $\int_0^1 x^n d\mu(x) = 0$, $n = 1, 2, \ldots$. Since we know $\int_0^1 d\mu(x) = 0$ we have that $d\mu(x) \equiv 0$. ∎

As another example of the techniques involved in proving completeness we have the following:

THEOREM 8. If $\{\lambda_n\}$ are bounded away from $[-1, 0]$ then $\{1/(x+\lambda_n)\}$ is complete in $C[0, 1]$.

Proof. Assume $\int_0^1 d\mu(x)/(x + \lambda_n) = 0$ and let

$$f(z) = \int_0^1 \frac{d\mu(x)}{x + z}.$$

Then $f(z)$ is analytic in the complement of $[-1, 0]$ and since, as $|z|$ approaches infinity, $f(z)$ approaches 0, $f(z)$ is analytic at infinity also.

We now consider two cases:

(a) If $|\lambda_n| < M$ for some M and all n, then the λ_n have a convergent subsequence which converges to a point outside of $[-1, 0]$. Since $f(\lambda_n) = 0$ we would get that $f(z) \equiv 0$.

(b) If there is a subsequence $\{\lambda_{n_k}\}$ such that $|\lambda_{n_k}|$ approaches infinity, then, since $f(z)$ is analytic at infinity and its zeroes accumulate there, we will have $f(z) \equiv 0$.

Therefore, in both cases,

$$\int_0^1 \frac{d\mu(x)}{x + z} \equiv 0.$$

Then

$$\int_{|z|=2} e^{2\pi i k z} \int_0^1 \frac{d\mu(x)}{x + z} dz = 0.$$

However, we can switch the order of integration and get

$$0 = \int_0^1 \left(\int_{|z|=2} \frac{e^{2\pi i k z}}{x+z} dz \right) d\mu(x).$$

By residues the inner integral is $e^{-2\pi i k x}$. Thus

$$\int_0^1 e^{2\pi i n x} d\mu(x) = 0 \quad n = 0, \pm 1, \pm 2, \ldots.$$

In addition, if we set $k = 1/2$, we have that

$$\int_0^1 e^{-\pi i x} d\mu(x) = 0.$$

Therefore, $d\mu(x) \equiv 0$ by the following theorem.

THEOREM 9. *Let $f \in C[0, 1]$ be such that $f(0) \neq f(1)$. Then*

$$\{e^{2\pi i n x}\}_{n=-\infty}^{\infty} \cup \{f(x)\}$$

is complete in $C[0, 1]$.

Proof. Let $g(x)$ be any continuous function on $[0, 1]$ and let

$$C = \frac{g(1) - g(0)}{f(1) - f(0)}.$$

Then $g(x) - Cf(x) \in C[0, 1]$ and $g(0) - Cf(0) = g(1) - Cf(1)$. Therefore, we can approximate $g(x) - Cf(x)$ by $\sum_{n=-N}^{N} a_n e^{2\pi i n x}$. Thus $g(x)$ can be approximated by a linear combination of

$$\{e^{2\pi i n x}\}_{n=-\infty}^{\infty} \cup \{f(x)\}. \quad \blacksquare$$

(4) Codimension

The fact that $\{e^{2\pi i n x}\}_{n=-\infty}^{\infty}$ is incomplete in $C[0, 1]$ but can be complete with the addition of one function can be expressed as "$\{e^{2\pi i n x}\}_{n=-\infty}^{\infty}$ has codimension 1 in $C[0, 1]$." For that, however, we need the following definition.

Definition. A set of vectors $\{V_k\}$ in a Banach space B is said to have codimension n in B if there exist $f_1, \ldots, f_n \in B$ such that

$$\{f_1, \ldots, f_n\} \cup \{V_k\}$$

is complete in B but $\{g_1, \ldots, g_{n-1}\} \cup \{V_k\}$ is incomplete in B for any $g_1, \ldots, g_{n-1} \in B$.

The next theorem provides, what turns out to be, in practice, a very useful way of finding lower bounds for the codimension.

THEOREM 10. $\{V_k\}$ has codimension n iff the span of the space of linear functionals on B which are orthogonal to all V_k, is n.

Proof. Assume the codimension is n and assume there are $n+1$ linear functionals $L_1, \ldots, L_{n+1}$ such that $L_i(V_k) = 0, i = 1, \ldots, n+1$.

There exist $f_1, \ldots, f_n \in B$ such that $\{f_1, \ldots, f_n\} \cup \{V_k\}$ is complete in B. Consider $C_1 L_1 + \cdots + C_{n+1} L_{n+1} = L_0$ where $C_1, \ldots, C_{n+1}$ are chosen so that $L_0(f_i) = 0$ for $i = 1, \ldots, n$. We can find such $C_1, \ldots, C_{n+1}$ since there are more unknowns than equations. Obviously $L_0(V_k) = 0$. Since $\{f_1, \ldots, f_n\} \cup \{V_k\}$ is complete in B, we must have $L_0 \equiv 0$. Therefore the codimension is greater than or equal to the number of linearly independent linear functionals orthogonal to $\{V_k\}$.

We now prove the existence of n linearly independent linear functionals orthogonal to $\{V_k\}$. Since $\{V_k\}$ is incomplete, there is a L_1 such that $L_1(V_k) = 0$ and $L_1 \not\equiv 0$; i.e., there is an $f_1 \in B$ such that $L_1(f_1) \not\equiv 0$. If $n = 1$ we are done. If not, then $\{f_1\} \cup \{V_k\}$ is incomplete in B; i.e., there is an L_2 and $f_2 \in B$ such that $L_2(V_k) = 0 = L_2(f_1)$ but $L_2(f_2) \not\equiv 0$. We keep on doing this and get $L_1, \ldots, L_n$ and $f_1, \ldots, f_n$ such that $L_i(V_k) = 0, i = 1, 2, \ldots, n$ and $L_i(f_j) = 0, j < i$ but $L_i(f_i) \not\equiv 0$.

We will now prove the $\{L_i\}$ are linearly independent. Assume

$$C_1 L_1 + \cdots + C_n L_n = 0.$$

Applying that linear functional in turn to $f_1, f_2, \ldots, f_n$, we get that $C_1, \ldots, C_n$ are zero. Thus the number of linearly independent linear functionals orthogonal to $\{V_k\}$ is greater or equal to the codimension and hence equal to it. ∎

We now apply this theorem to questions of completeness of various sequences of functions in different Banach spaces. We know that $\{e^{inx}\}_{n=-\infty}^{\infty}$ has codimension 1 in $C[-\pi, \pi]$. (This follows by replacing x by $2\pi x$ and using theorem 9.) However, in $L^p[-\pi, \pi]$, it is a different story.

THEOREM 11. $\{e^{inx}\}_{n=-\infty}^{\infty}$ is complete in $L^P[-\pi, \pi], 1 \leq P < \infty$.

Proof. Let $d\mu(x)$ have a mass point of -1 at $-\pi$ and 1 at π. Then

$$\int_{-\pi}^{\pi} e^{inx} d\mu(x) = 0 \quad n = 0, \pm 1, \pm 2 \ldots$$

This $d\mu(x)$ is not a linear functional on $L^P[-\pi, \pi]$. If $\{e^{inx}\}_{-\infty}^{\infty}$ were incomplete, then there would be an $f(x) \in L^q[-\pi, \pi]$ (where $q = P/(P-1)$ if $P \neq 1$, and $q = \infty$ if $P = 1$) such that

$$\int_{-\pi}^{\pi} e^{inx} f(x) dx = 0 \quad n = 0, \pm 1, \pm 2 \ldots.$$

Let $d\mu(x) = f(x)\,dx$. Then $d\mu_1(x)$ is a finite measure on $[-\pi, \pi]$ such that $\int_{-\pi}^{\pi} e^{inx} d\mu_1(x) = 0$, $n = 0, \pm 1, \pm 2 \ldots$ and $d\mu_1(x)$ gives us a linear functional on $C[-\pi, \pi]$. Since the codimension of $\{e^{inx}\}_{-\infty}^{\infty}$ in $C[-\pi, \pi]$ is 1 then $d\mu_1(x) = C d\mu(x)$. However, unless $C = 0$, this is impossible as $d\mu_1(x)$ is absolutely continuous and $C d\mu(x)$ is not. Thus, $f(x) = 0$ a.e. ∎

THEOREM 12. $\{\cos nx\}_{n=1}^{\infty}$ is incomplete in $L^1[0, \pi]$ (and hence in $L^P[0, \pi]$, $P \geq 1$ and in $C[0, \pi]$).

Proof. $\int_0^{\pi} \cos nx\, dx = 0$, $n = 1, 2 \ldots$, and the constant function is in $L^{\infty}[0, \pi]$. ∎

THEOREM 13. $\{\sin nx\}_{n=1}^{\infty}$ is complete in $L^P[0, \pi]$, $1 \leq P < \infty$.

Proof. Take any $f \in L^P[0, \pi]$ and extend it to $[-\pi, 0]$ by defining $f(-x) = -f(x)$. Then $f \in L^P[-\pi, \pi]$ and f is odd. Since $\{e^{inx}\}_{-\infty}^{\infty}$ is complete in $L^P[-\pi, \pi]$, we can approximate f in $L^P[-\pi, \pi]$ by

$$\sum_{-N}^{N} a_n e^{inx} = a_0 + \sum_{n=1}^{N} (a_n + a_{-n}) \cos nx + i \sum_{n=1}^{N} (a_n - a_{-n}) \sin nx$$

However, since $f(x)$ is odd, we can easily show that

$$a_0 + \sum_{n=1}^{N} (a_n + a_n) \cos nx$$

hurts the approximation.

Hence $f(x)$ can be approximated in $L^P[-\pi, \pi]$ (and certainly in $L^P[0, \pi]$) by $\{\sin nx\}_{n=1}^{\infty}$. ∎

THEOREM 14. $\{\cos nx\}_{n=0}^{\infty}$ is complete in $C[0, \pi]$ (and hence in $L^P[0, \pi]$, $1 < P < \infty$).

Proof. Assume there is a $d\mu_1(x)$ such that

$$\int_0^{\pi} \cos nx\, d\mu_1(x) = 0, \quad n = 0, 1, 2 \ldots.$$

Extend $\mu_1(x)$ to $[-\pi, 0]$ by $\mu_1(-x) = \mu_1(x)$. Then

$$\int_{-\pi}^{\pi} \cos nx\, d\mu_1(x) = 0 = \int_{-\pi}^{\pi} \sin nx\, d\mu_1(x).$$

Hence $d\mu_1(x)$ is orthogonal to $\{e^{inx}\}_{-\infty}^{\infty}$. Since $\{e^{inx}\}_{-\infty}^{\infty}$ has codimension 1 in $C[-\pi, \pi]$, $d\mu_1(x)$ must be equal to $C\, d\mu(x)$ where $d\mu(x)$ has a mass of -1 at $-\pi$ and 1 at π. However, $0 = \int_0^{\pi} d\mu_1(x) = \int_0^{\pi} C\, d\mu(x) = C$ and hence $d\mu_1(x) \equiv 0$. ∎

We thus see that $\{\cos nx\}_{n=1}^{\infty}$ has codimension 1 in $C[0, \pi]$ and

$L^P[0, \pi]$, $1 \leq P < \infty$.

THEOREM 15. $\{1\} \cup \{\sin nx\}_{n=1}^{\infty}$ is incomplete in $C[0, \pi]$.

Proof. Let $d\mu_1(x)$ have a mass of 1 at 0 and a mass of -1 at π. Then

$$\int_0^\pi \sin nx \, d\mu_1(x) = \int_0^\pi d\mu_1(x) = 0.$$

What is the codimension of $\{\sin nx\}_{n=1}^{\infty}$ in $C[0, \pi]$? Let $d\mu_2(x)$ have a mass of 1 at 0 and 1 at π. Then

$$\int_0^\pi \sin nx \, d\mu_2(x) = 0, \quad n = 1, 2 \ldots,$$

and obviously $d\mu_1(x)$ and $d\mu_2(x)$ are linearly independent. Thus, the codimension of $\{\sin nx\}_{n=1}^{\infty}$ is ≥ 2 in $C[0, \pi]$. On the other hand, $\{1, x\} \cup \{\sin nx\}_{n=1}^{\infty}$ is complete in $C[0, \pi]$ by the same method that we proved $\{e^{inx}\}_{n=-\infty}^{\infty} \cup \{f(x)\}$ is complete in $C[-\pi, \pi]$ if $f(-\pi) \neq f(\pi)$. Thus $\{\sin nx\}_{n=1}^{\infty}$, while complete in $L^P[0, \pi]$ for $1 \leq P < \infty$, has codimension 2 in $C[0, \pi]$.

CHAPTER XI

A MÜNTZ-JACKSON THEOREM

In Chapter X, we proved Müntz's theorem. That told us when $\{x^{\lambda_i}\}_{i=0}^{\infty}$ is complete in $C[0, 1]$. However, following the general motif of most of this book, the natural question to ask is "How well do they approximate?"

On becoming slightly more specific, we realize the question can be rephrased as "Given $\lambda_0 = 0$, positive numbers $\lambda_1, \ldots, \lambda_N$ and $f \in C[0, 1]$, how closely can functions of the form $\sum_{k=0}^{N} c_k x^{\lambda_k}$ approximate $f(x)$?"

To get an idea of what kinds of theorems to expect we recall the case $\lambda_i = i$. There we saw two important theorems. First we saw that there is an $A > 0$ such that for every $f \in C[0, 1]$ and every positive integer N, we could find $c_0, c_1, \ldots, c_N$ such that $\|f - \sum_{i=0}^{N} c_i x^i\| \leq A\omega(f, 1/N)$.

We also saw that there is a $B > 0$ such that for each positive integer N there is an $f \in C[0, 1]$ such that $\|f - \sum_{i=0}^{N} c_i x^i\| > B\omega(f, 1/N)$ for any $c_0, c_1, \ldots, c_N$. We thus say that $1/N$ is the correct "approximation index" for $\{x^i\}_{i=0}^{N}$.

We would like to find similar theorems for the general case. That is, we would like to find an ε_Λ, which is a function of the set $\Lambda = \{\lambda_0, \lambda_1, \lambda_2, \ldots, \lambda_N\}$ such that

(1) There is an $A > 0$, such that for each $f \in C[0, 1]$ we can find $c_0, c_1, \ldots, c_N$ such that $\|f - \sum_{i=0}^{N} c_i x^{\lambda_i}\| \leq A\omega(f, \varepsilon_\Lambda)$.

(2) There is a $B > 0$, such that for each Λ, there is an $f \in C[0, 1]$ such that $\|f - \sum_{i=0}^{N} c_i x^{\lambda_i}\| > B\omega(f, \varepsilon_\Lambda)$ for any choice of $c_0, c_1, \ldots, c_N$. Such an ε_Λ would then be called the approximation index for the set $\{x^{\lambda_i}\}_{i=0}^{N}$.

Notation. 1) $E_\Lambda(f) = \inf_{c_0, c_1, \ldots, c_N} \|f - \sum_{i=0}^{N} c_i x^{\lambda_i}\|$
2) $\rho_\Lambda = \sup_{f \in \mathscr{F}} E_\Lambda(f)$.

As we shall see, once we know the growth of ρ_Λ, we'll know the approximation index for $\{x^{\lambda_i}\}_{i=0}^{N}$. First we'll need the following

THEOREM 1. For any $f \in C[0, 1]$, $E_\Lambda(f) \leq 2\omega(f, \rho_\Lambda)$.

Proof. For any $f \in C[0, 1]$, define $f(x)$, for $x \in [1, \rho_\Lambda]$ by $f(x) \equiv f(1)$. It is obvious that the modulus of continuity hasn't changed. Let

$$F(x) = \frac{1}{\rho_\Lambda} \int_x^{x+\rho_\Lambda} f(t)dt \ .$$

Then $F(x)$ is absolutely continuous and

$$\|F'\| = \left\|\frac{f(x + \rho_\Lambda) - f(x)}{\rho_\Lambda}\right\| \leq \frac{\omega(f, \rho_\Lambda)}{\rho_\Lambda} \ .$$

Thus

$$\frac{F(x)\rho_\Lambda}{\omega(f, \rho_\Lambda)} \in \mathscr{S}$$

by the following easily proven (as an exercise)

THEOREM 2. $f \in \mathscr{S}$ iff f is absolutely continuous and $|f'(x)| \leq 1$ a.e.

Consequently, we can find a $Q(x) = a_0 + a_1 x^{\lambda_1} + \cdots + a_N x^{\lambda_N}$ such that

$$\left\|\frac{F(x)\rho_\Lambda}{\omega(f, \rho_\Lambda)} - Q(x)\right\| \leq \rho_\Lambda$$

or $\|F(x) - P(x)\| \leq \omega(f, \rho_\Lambda)$ (where $P(x) = \omega(f, \rho_\Lambda) Q(x)/\rho_\Lambda$). Also

$$\|f(x) - F(x)\| = \left\|\frac{1}{\rho_\Lambda}\int_x^{x+\rho_\Lambda}(f(x) - f(t))dt\right\|$$

$$\leq \frac{1}{\rho_\Lambda}\int_x^{x+\rho_\Lambda}\|f(x) - f(t)\|dt$$

$$\leq \omega(f, \rho_\Lambda) \ .$$

Thus

$$E_\Lambda(f) \leq \|f(x) - P(x)\| \leq \|f(x) - F(x)\| + \|F(x) - P(x)\|$$
$$\leq 2\omega(f, \rho_\Lambda). \ \blacksquare$$

The essence of this chapter will be the proving of the inequalities $(1/48)\varepsilon_\Lambda \leq \rho_\Lambda \leq 184\varepsilon_\Lambda$ (where we obviously have not attempted to get best constants, just order of magnitude of ρ_Λ) where $\varepsilon_\Lambda = \sup_{\operatorname{Re} z = 1} |B(z)/z|$ and $B(z)$ is the Blaschke product $\prod_{i=1}^N (z - \lambda_i)/(z + \lambda_i)$. Then, by the previous theorem, we'll have $E_\Lambda(f) \leq 2\omega(f, 184\varepsilon_\Lambda) \leq 368\omega(f, \varepsilon_\Lambda)$. Also, since $\rho_\Lambda \geq (1/48)\varepsilon_\Lambda$, for each Λ we could find an $f \in \mathscr{S}$ such that $E_\Lambda(f) \geq (1/50)\varepsilon_\Lambda$. Then, since $\omega(f, \delta) \leq \delta$ we would have

$E_\Lambda(f) \geq (1/50)\omega(f, \varepsilon_\Lambda)$ and would have established ε_Λ as the approximation index for $\{x^{\lambda_i}\}_{i=0}^N$.

Before we start to prove the aforementioned inequalities, we will need two preliminary theorems.

Notation. By $c(\Lambda)$ we mean all $\mu(x)$ which are of bounded variation on $[0, 1]$ such that
(1) $\int_0^1 x^{\lambda_i} d\mu(x) = 0$, $i = 0, \ldots, N$.
(2) $\int_0^1 |d\mu(x)| \leq 1$.
(3) $\mu(1) = 0$.

THEOREM 3. $\rho_\Lambda = \sup_{\mu \in c(\Lambda)} \int_0^1 |\mu(x)| dx$.

Proof. Take any $f \in \mathscr{S}$ and $\mu \in c(\Lambda)$.
Then $\int_0^1 f(x) d\mu(x) = -\int_0^1 f'(x) \mu(x) dx$ (since $\int_0^1 d\mu(x) = 0$ and $\mu(1) = 0$ we get $\mu(0) = 0$). However, for this $f(x)$, we find $a_0, \ldots, a_N$ such that $\|f(x) - \sum_{n=0}^N a_n x^{\lambda_n}\| \leq \rho_\Lambda$. Therefore,

$$\left| \int_0^1 f'(x) \mu(x) dx \right| = \left| \int_0^1 f(x) d\mu(x) \right|$$
$$= \left| \int_0^1 \left(f(x) - \sum_{n=0}^N a_n x^{\lambda_n} \right) d\mu(x) \right|$$
$$\leq \rho_\Lambda \int_0^1 |d\mu(x)| \leq \rho_\Lambda.$$

If, instead of choosing $f(x)$ and $\mu(x)$ independently, we were to choose $f(x) \in \mathscr{S}$ such that $f'(x) = \operatorname{sgn} \mu(x)$ we would have $\int_0^1 |\mu(x)| dx \leq \rho_\Lambda$. Since $\mu(x)$ was arbitrary we get

$$\sup_{\mu \in c(\Lambda)} \int_0^1 |\mu(x)| dx \leq \rho_\Lambda.$$

To prove the inequality the other way, we taken an $\varepsilon > 0$ and find an $f \in \mathscr{S}$ such that $E_\Lambda(f) > \rho_\Lambda - \varepsilon$. By the Hahn-Banach theorem, we find a $\mu \in c(\Lambda)$ such that $\int_0^1 f(x) d\mu(x) = E_\Lambda(f)$. Therefore,

$$\rho_\Lambda - \varepsilon < \int_0^1 f(x) d\mu(x)$$
$$= \left| \int_0^1 f'(x) \mu(x) dx \right|$$
$$\leq \int_0^1 |\mu(x)| dx.$$

Thus,

$$\sup_{\mu \in c(\Lambda)} \int_0^1 |\mu(x)| dx > \rho_\Lambda - \varepsilon$$

and since ε was arbitrary,

$$\sup_{\mu \in c(\Lambda)} \int_0^1 |\mu(x)| dx \geq \rho_\Lambda$$

and hence is equal to it. ∎

If we make the change of variable $x = e^{-t}$, we get the following form of the previous

THEOREM 4. $\rho_\Lambda = \sup \int_0^\infty e^{-t} |\mu(t)| dt$ where the sup is taken over all μ such that
  (a) $\mu(0) = 0$,
  (b) $\int_0^\infty |d\mu(t)| \leq 1$,
  (c) $\int_0^\infty e^{-t\lambda_i} d\mu(t) = 0$, $i = 0, \ldots, N$.

(1) The Upper Bound

We take an arbitrary $f \in \mathscr{S}$ (that is $\mathscr{S}([0, 1])$) and we would like to get an upper bound for $E_\Lambda(f)$. We first define $f_1(x) = f(x + 1/2)$. Then $f_1 \in \mathscr{S}([-1/2, 1/2])$. We extend f_1 to be in $\mathscr{S}([-1, 1])$. We then let $g(\theta) = f_1(\cos \theta)$ which is in $\mathscr{S}^*$. If, for any M, we take the trigonometric polynomial T of degree M determined from Jackson's theorem (i.e., $T = g * K_M$), we know that

$$\|g - T\| \leq \frac{17}{M}.$$

Moreover, since $T(\theta)$ is an even function of θ, $T(\text{arc cos } x) = P_1(x)$ is in $\mathscr{S}_M$ and

$$\max_{x \in [-1,1]} |g(\text{arc cos } x) - T(\text{arc cos } x)| \leq \frac{17}{M}$$

or

$$\max_{x \in [-1,1]} |f_1(x) - P_1(x)| \leq \frac{17}{M}.$$

We will now get a bound on $|P_1'(x)|$.

$$\frac{d}{dx} P_1(x) = \frac{dT}{dx}(\text{arc cos } x) = \frac{dT(\text{arc cos } x)}{d(\text{arc cos } x)} \frac{d(\text{arc cos } x)}{dx}$$
$$= \frac{d(T(\theta))}{d\theta} \left(\frac{-1}{\sqrt{1-x^2}}\right).$$

We thus need to estimate $|dT(\theta)/d\theta|$. We recall that $T(\theta) = (g * K_M)(\theta)$. Then

$$T' = (g*K_M)'(\theta)$$
$$= (g'*K_M)(\theta)$$

and

$$|T'(\theta)| = \left|\int_{-\pi}^{\pi} g'(t-\theta)K_M(t)dt\right|$$
$$\leq \int_{-\pi}^{\pi} |g'(t-\theta)||K_M(t)|dt$$
$$\leq \int_{-\pi}^{\pi} |K_M(t)|dt \quad (g \text{ is in } \mathscr{S}^*)$$
$$= 1.$$

Thus,

$$|P_1'(x)| \leq \left|\frac{1}{\sqrt{1-x^2}}\right|$$

which is less than 2 for $x \in [-1/2, 1/2]$.

Summarizing, we have, for any $x \in [-1/2, 1/2]$, $|f_1(x) - P_1(x)| \leq 17/M$ and $|P_1'(x)| \leq 2$. If we now shift back to $[0, 1]$ and let $P(x) = P_1(x-1/2)$, we have $\|f(x) - P(x)\| \leq 17/M$ and $\|P'\| \leq 2$. By applying theorem 2.12 to the polynomial $P'(x)$, we get $|P^{(k)}(0)| \leq 2(M-1)^{k-1} < 2M^{k-1}$.

Instead of estimating $E_\Lambda(f)$ we'll estimate $E_\Lambda(P)$ and combine that with $\|f - P\| \leq 17/M$ to get an estimate on $E_\Lambda(f)$.

Since $P(x) = P(0) + P'(0)x + \cdots + (P^{(M)}(0)/M!)x^M$ we have that

$$E_\Lambda(P) \leq \sum_{k=0}^{M} \frac{|P^{(k)}(0)|}{k!} E_\Lambda(x^k) \leq \sum_{k=0}^{M} \frac{2M^{k-1}}{k!} E_\Lambda(x^k).$$

We thus need an estimate on $E_\Lambda(x^k)$. For $k=0$ this is 0, whereas for $k = 1, \ldots M$ this will be provided by the next two lemmas.

LEMMA 5. $E_\Lambda(x^k) \leq |B(k)| \quad k = 1, 2, \ldots$.

Proof. By the Hahn-Banach theorem, we can find a $\mu \in c(\Lambda)$ such that $\int_0^1 x^k d\mu(x) = E_\Lambda(x^k)$. If we look at $\int_0^1 x^z d\mu(x)$ we see that it is a bounded analytic function in $\operatorname{Re} z > 0$ which vanishes for $z \in \Lambda$. Thus $\int_0^1 x^z d\mu(x) = F(z)B(z)$ where $F(z)$ is bounded and analytic in $\operatorname{Re} z > 0$. For $\operatorname{Re} z = 0$ we have $|F(z)| = |F(z)B(z)| \leq \int_0^1 |d\mu(x)| \leq 1$ and therefore $|F(z)| \leq 1$ in $\operatorname{Re} z \geq 0$. In particular, $|\int_0^1 x^k d\mu(x)| = |F(k)B(k)| \leq |B(k)|$ and thus $E_\Lambda(x^k) \leq |B(k)|$. ∎

LEMMA 6. $|B(k)| \leq (\varepsilon_\Lambda k)^k, \quad k = 1, 2, \ldots$.

Proof. We need to show that for some real y,

$$\prod_{i=1}^{N} \left| \frac{k - \lambda_i}{k + \lambda_i} \right| \leq \left| \frac{k}{1 + iy} \prod_{j=1}^{N} \frac{1 + iy - \lambda_j}{1 + iy + \lambda_j} \right|^k.$$

We will prove it for $y = \sqrt{k^2 - 1}$. For that y the inequality is (after squaring)

$$\prod_{i=1}^{N} \left(\frac{k - \lambda_i}{k + \lambda_i} \right)^2 \leq \prod_{i=1}^{N} \left(\frac{k^2 + \lambda_i^2 - 2\lambda_i}{k^2 + \lambda_i^2 + 2\lambda_i} \right)^k$$

and we will see that this holds, factor by factor. The elementary inequality $(1 - kt)/(1 + kt) \leq ((1 - t)/(1 + t))^k$ (for $t \in [0, 1]$) is easily proven by induction and then by setting $t = 2\lambda_i/(\lambda_i^2 + k^2)$ we get the desired inequality. ∎

We are now able to further our estimate on $E_\Lambda(P)$. We had before

$$E_\Lambda(P) \leq \sum_{k=0}^{M} \frac{2M^{k-1}}{k!} E_\Lambda(x^k)$$

which we now know is less than or equal to

$$\sum_{k=1}^{M} \frac{2M^{k-1}}{k!} (\varepsilon_\Lambda k)^k = \frac{2}{M} \sum_{k=1}^{M} \frac{(M\varepsilon_\Lambda k)^k}{k!}$$

$$\leq \frac{2}{M} \sum_{k=1}^{M} (e\varepsilon_\Lambda M)^k$$

where we used the inequality $k^k/k! \leq e^k$.

Since $\|f - P\| \leq 17/M$ we easily get (by the triangle inequality) that $E_\Lambda(f) \leq 17/M + E_\Lambda(P) \leq 17/M + (2/M) \sum_{k=1}^{M} (e\varepsilon_\Lambda M)^k$. Until now M was arbitrary. If $1/8\varepsilon_\Lambda \leq 1$ we choose $M = 1$ and get an estimate on $E_\Lambda(f)$ of $17 + 2e\varepsilon_\Lambda$ which is smaller than $142\varepsilon_\Lambda$.

If $1/8\varepsilon_\Lambda > 1$, we let $M = [1/4\varepsilon_\Lambda]$. Thus $M \leq 1/4\varepsilon_\Lambda$ and $1/M \leq 8\varepsilon_\Lambda$. We then get an estimate for $E_\Lambda(f)$ of $17(8\varepsilon_\Lambda) + 2(8\varepsilon_\Lambda)e/(4-e)$ which is smaller than $184\varepsilon_\Lambda$. Thus in any case, $E_\Lambda(f) \leq 184\varepsilon_\Lambda$. Since f was an arbitrary function in $\mathscr{S}$ and since $\rho_\Lambda = \sup_{f \in \mathscr{S}} E_\Lambda(f)$ we have proven the following

THEOREM 7. $\rho_\Lambda \leq 184\varepsilon_\Lambda$.

(2) The Lower Bound

We are now going to prove the lower bound part of the inequality for ρ_Λ; that is $\rho_\Lambda \geq \varepsilon_\Lambda/48$.

We recall that by theorem 4, $\rho_\Lambda = \sup \int_0^\infty e^{-t} |\mu(t)| dt$ where the sup is taken over all $\mu(x)$ such that

(a) $\mu(0) = 0$.

(b) $\int_0^\infty |d\mu(x)| \leq 1$.
(c) $F(z) = \int_0^\infty e^{-zt} d\mu(t)$ vanishes at $z = \lambda_n$, $n = 0, \ldots, N$.

Thus, it will suffice to find a $\mu(x)$ which satisfies our conditions and in addition has $\int_0^\infty e^{-t} |\mu(t)|\, dt > \varepsilon_\Lambda/48$.

We remember that $\varepsilon_\Lambda = \max_{\text{Re}\, z=1} |B(z)/z|$ and we let $1 + i\tau$ be a point where that maximum is achieved.

Motivated in part by c) we consider

$$F(z) = \frac{4z^2 s^2 B(z)}{(z+s)^4 (z+1-i\tau)}$$

where $s = \sqrt{1+\tau^2}$. We will prove that this function can be expressed in the form $\int_0^\infty e^{-zt} h(t)\, dt$ where $h(t) \in L^1[0, \infty)$. If we set $\mu(x) = \int_0^x h(t)\, dt$, we will get $d\mu(x) = h(x)\, dx$ and we will have gotten $F(z)$ in the form of part c) and thereby satisfy part c) (since $F(z)$ vanishes at λ_n, $n = 0, \ldots N$). Part a) will obviously be satisfied leaving us just to prove

$$\frac{\int_0^\infty e^{-t} \left| \int_0^t h(x)\, dx \right| dt}{\int_0^\infty |h(t)|\, dt} > \frac{\varepsilon_\Lambda}{48}.$$

(The denominator comes in because we want to satisfy condition b).)

Notation. For the rest of this chapter, all norms will refer to $[0, \infty)$. This is,

$$\|f\| = \sup_{x \in [0,\infty)} |f(x)|,$$

$$\|f\|_{L^1} = \int_0^\infty |f(x)|\, dx$$

and

$$\|f\|_{L^2} = \left(\int_0^\infty |f(x)|^2\, dx \right)^{1/2}.$$

Thus, what we have to do, is to prove $F(z) = \int_0^\infty e^{-zt} h(t)\, dt$ where $h \in L^1[0, \infty)$ and then prove

$$\frac{\left\| e^{-t} \int_0^t h(x)\, dx \right\|_{L^1}}{\|h\|_{L^1}} > \frac{\varepsilon_\Lambda}{48}.$$

We fix an $x > 0$ and consider

$$F(x + iy) = \frac{4(x+iy)^2 s^2 B(x+iy)}{(x+iy+s)^4 (x+iy+1-i\tau)}$$

as a function of y.

Since $|B(z)| \leq 1$ in $\text{Re } z \geq 0$, we easily get that $\int_{-\infty}^{\infty} |F(x + iy)|^2 dy$ is uniformly bounded in $x > 0$. Since $F(z)$ is analytic in $\text{Re } z > 0$, we get that $F(z) = \int_0^\infty e^{-tz} h(t) dt$ where $h \in L^2[0, \infty)$ by the

Paley Wiener Theorem: $F(z)$ is analytic in $\text{Re } z > 0$ and

$$\int_{-\infty}^{\infty} |F(x + iy)|^2 dy$$

is uniformly bounded in $x > 0$ iff $F(z) = \int_0^\infty e^{-zt} g(t) dt$ where $g \in L^2[0, \infty)$.

We notice that $F(iy) = \int_0^\infty e^{-iyt} h(t) dt$ and thus $F(iy)$ is the Fourier transform of $h(t)$.

We also notice that if we perform an integration by parts on $\int_0^\infty e^{-zt} h(t) dt$ we get

$$F(z) = -z \int_0^\infty e^{-zt} \left[\int_0^t h(x) dx \right] dt$$

and in particular,

$$\frac{F(1 + i\tau)}{1 + i\tau} = -\int_0^\infty e^{-(1+i\tau)t} \left[\int_0^t h(x) dx \right] dt$$

THEOREM 8. $\left\| e^{-t} \int_0^t h(x) dx \right\|_{L^1} \geq \varepsilon_\Lambda / 8$.

Proof.

$$\left\| e^{-t} \int_0^t h(x) dx \right\|_{L^1} = \int_0^\infty e^{-t} \left| \int_0^t h(x) dx \right| dt$$

$$= \int_0^\infty \left| e^{-(1+i\tau)t} \int_0^t h(x) dx \right| dt$$

$$\geq \left| \int_0^\infty e^{-(1+i\tau)t} \int_0^t h(x) dx \, dt \right|$$

$$= \left| \frac{F(1 + i\tau)}{1 + i\tau} \right|$$

$$= 2s^2 \left| \frac{(1 + i\tau)^2}{(1 + i\tau + s)^4} \right| \left| \frac{B(1 + i\tau)}{(1 + i\tau)} \right|$$

$$= 2s^2 \varepsilon_\Lambda \left| \frac{(1 + i\tau)^2}{(1 + i\tau + s)^4} \right|$$

$$= \frac{s^2}{2(1 + s)^2} \varepsilon_\Lambda$$

$$\geq \frac{\varepsilon_\Lambda}{8} \text{ (since } s \geq 1\text{)}. \blacksquare$$

To complete our estimating, we need only find an upper bound for $\|h\|_{L^1}$. To do that we will need the following lemma (due to Carlson).

LEMMA 9. Let $g(y) = \int_{-\infty}^{\infty} e^{-ixy} G(x) dx$ where g, g′ and G are all in $L^2(-\infty, \infty)$. Then $(\int_{-\infty}^{\infty} |G(x)| dx)^4 \leq (\int_{-\infty}^{\infty} |g(x)|^2 dx)(\int_{-\infty}^{\infty} |g'(x)|^2 dx.)$

Proof. By Schwarz's inequality,

$$\left(\int_{-\infty}^{\infty} |G(x)| dx\right)^2 \leq \int_{-\infty}^{\infty} \frac{dx}{a^2 + x^2} \int_{-\infty}^{\infty} |G(x)|^2 (a^2 + x^2) dx$$

$$= \frac{\pi}{a}\left[a^2 \int_{-\infty}^{\infty} |G(x)|^2 dx + \int_{-\infty}^{\infty} |G(x)x|^2 dx\right]$$

$$= \frac{\pi}{a}\left[\frac{a^2}{2\pi}\int_{-\infty}^{\infty} |g(x)|^2 dx + \frac{1}{2\pi}\int_{-\infty}^{\infty} |g'(x)|^2 dx\right]$$

where we have used Parseval's identity in the last line. If we now set

$$a = \left(\frac{\int_{-\infty}^{\infty} |g'(x)|^2 dx}{\int_{-\infty}^{\infty} |g(x)|^2 dx}\right)^{1/2}$$

we have proven the lemma. ∎

We will apply this lemma in the following manner. We have that $F(iy) = \int_0^{\infty} e^{-iyt} h(t) dt$. We define $h(t) = 0$ for $t < 0$ and then we have $F(iy) = \int_{-\infty}^{\infty} e^{-iyt} h(t) dt$. Thus the lemma would tell us that

$$\|h\|_{L^1}^4 = \left(\int_0^{\infty} |h(x)| dx\right)^4 = \left(\int_{-\infty}^{\infty} |h(x)| dx\right)^4$$

$$\leq \left(\int_{-\infty}^{\infty} |F(iy)|^2 dy\right)\left(\int_{-\infty}^{\infty} |F'(iy)|^2 dy\right)$$

However, it turns out that it is much simpler to consider $F(z)e^{-cz}$ instead of $F(z)$ where

$$c = 2\sum_{i=1}^{N} \frac{\lambda_i}{\lambda_i^2 + s^2}.$$

We have, first of all, that

$$F(z)e^{-cz} = \int_0^{\infty} e^{-z(t+c)} h(t) dt$$

$$= \int_{-\infty}^{\infty} e^{-z(t+c)} h(t) dt$$

$$= \int_{-\infty}^{\infty} e^{-zt} h(t-c) dt.$$

Therefore, $F(iy) e^{-c(iy)} = \int_{-\infty}^{\infty} e^{-iyt} h(t - c) dt$ and the lemma then tells us that

$$\|h\|_{L^1}^4 = \left(\int_{-\infty}^{\infty} |h(x)| dx\right)^4 = \left(\int_{-\infty}^{\infty} |h(x - c)| dx\right)^4$$

$$\leq \left(\int_{-\infty}^{\infty} |F(iy) e^{-ciy}|^2 dy\right)\left(\int_{-\infty}^{\infty} |(F(iy)e^{-ciy})'|^2 dy\right)$$

$$= \left(\int_{-\infty}^{\infty} |F(iy)|^2 dy\right)\left(\int_{-\infty}^{\infty} |F'(iy) - cF(iy)|^2 dy\right).$$

We first get an estimate on $\int_{-\infty}^{\infty} |F(iy)|^2 dy$. We have

$$|F(iy)| = \left|\frac{4(iy)^2 s^2 B(iy)}{(iy + s)^4 (iy + 1 - i\tau)}\right|$$

$$\leq \frac{4y^2 s^2}{(y^2 + s^2)^2} \frac{1}{|1 + i(y - \tau)|}$$

$$\leq \frac{1}{|1 + i(y - \tau)|}$$

so that

$$\int_{-\infty}^{\infty} |F(iy)|^2 dy \leq \int_{-\infty}^{\infty} \frac{dy}{1 + (y - \tau)^2}$$

$$= \pi.$$

To complete our estimating we just have to get an upper bound on $\int_{-\infty}^{\infty} |F'(iy) - cF(iy)|^2 dy$. In doing that we will need the following lemma which we'll prove later.

LEMMA 10. $|B'(iy) - cB(iy)| \leq (y^2 + s^2)^2/2y^2 s^2$

Now, by direct differentiation, we obtain

$$F'(z) - cF(z) = \frac{4z^2 s^2 B(z)}{(z + s)^4 (z + 1 - i\tau)}$$

$$\times \left[\frac{2}{z} - \frac{4}{z + s} - \frac{1}{z + 1 - i\tau} + \frac{B'(z)}{B(z)} - c\right].$$

Therefore,

$$|1 + i(y - \tau)||F'(iy) - cF(iy)|$$

$$= \frac{4y^2 s^2}{(y^2 + s^2)^2} \left|\frac{2}{iy} - \frac{4}{iy + s} - \frac{1}{1 + i(y - \tau)} + \frac{B'(iy)}{B(iy)} - c\right|$$

$$\leq \frac{8ys^2}{(y^2 + s^2)^2} + \frac{16y^2 s}{(y^2 + s^2)^2} + \frac{4y^2 s^2}{(y^2 + s^2)^2} + \frac{4y^2 s^2}{(y^2 + s^2)^2}|B'(iy) - cB(iy)|.$$

By lemma 10, the last term is ≤ 2. The first term is
$$\frac{2ys}{y^2+s^2} \cdot \frac{4s}{y^2+s^2} \leq \frac{2ys}{y^2+s^2} \cdot \frac{4}{s} \leq 4$$
since $s \geq 1$. The second term is
$$\frac{4y^2s^2}{(y^2+s^2)^2} \cdot \frac{4}{s} \leq 4$$
again since $s \geq 1$ and the third term is obviously ≤ 1. Altogether, we obtain
$$|F'(iy) - cF(iy)| \leq \frac{11}{|1+i(y-\tau)|}$$
and easily conclude that
$$\int_{-\infty}^{\infty} |F'(iy) - cF(iy)|^2 dy \leq 121\pi .$$
We thus have an over all estimate on $\|h\|_{L^1}$ and have proven:

THEOREM 11. $\|h\|_{L^1} \leq \sqrt{11\pi} < 6$.

Combining theorem 8 with theorem 11, we have:

THEOREM 12. $\rho_\Lambda > \varepsilon_\Lambda / 48$.

All we have to do now is the

Proof of lemma 10. We first consider $|B(1+iy)/(1+iy)|^2$ which is equal to
$$\frac{1}{1+y^2} \prod_{i=1}^{N} \frac{y^2 + (\lambda_i - 1)^2}{y^2 + (\lambda_i + 1)^2}$$
This expression is maximized by $y = \tau$. Therefore, the logarithmic derivative of this expression with respect to y^2,
$$4 \sum_{i=1}^{N} \frac{\lambda_i}{(y^2 + (\lambda_i - 1)^2)(y^2 + (\lambda_i + 1)^2)} - \frac{1}{y^2+1}$$
must be 0 at τ if $\tau \neq 0$ and must be ≤ 0 at τ if $\tau = 0$.

In either case, we have
$$4 \sum_{i=1}^{N} \frac{\lambda_i}{(\tau^2 + (\lambda_i - 1)^2)(\tau^2 + (\lambda_i + 1)^2)} \leq \frac{1}{\tau^2+1}$$
Since

$$(\tau^2 + (\lambda_i - 1)^2)(\tau^2 + (\lambda_i + 1)^2) = (\tau^2 + \lambda_i^2 + 1)^2 - 4\lambda_i^2$$
$$\leq (\tau^2 + \lambda_i^2 + 1)^2$$
$$= (s^2 + \lambda_i^2)^2$$

we get that

$$\sum_{i=1}^{N} \frac{\lambda_i}{(s^2 + \lambda_i^2)^2} \leq \frac{1}{4s^2}.$$

We now consider

$|B'(iy) - cB(iy)|$.

$$|B'(iy) - cB(iy)| = \left| \frac{B'(iy)}{B(iy)} - c \right|$$

$$= \left| 2 \sum_{i=1}^{N} \frac{\lambda_i}{\lambda_i^2 + y^2} - 2 \sum_{i=1}^{N} \frac{\lambda_i}{\lambda_i^2 + s^2} \right|$$

$$= 2(s^2 - y^2) \sum_{i=1}^{N} \frac{\lambda_i}{(\lambda_i^2 + y^2)(\lambda_i^2 + s^2)}$$

$$\leq 2(s^2 + y^2) \sum_{i=1}^{N} \frac{\lambda_i}{(\lambda_i^2 + s^2)^2} \left(1 + \frac{s^2 - y^2}{\lambda_i^2 + y^2} \right)$$

$$\leq 2(s^2 + y^2) \sum_{i=1}^{N} \frac{\lambda_i}{(\lambda_i^2 + s^2)^2} \left(1 + \frac{s^2}{\lambda_i^2 + y^2} \right)$$

$$\leq 2 \frac{(s^2 + y^2)^2}{y^2} \sum_{i=1}^{N} \frac{\lambda_i}{(\lambda_i^2 + s^2)^2}$$

$$\leq 2 \frac{(s^2 + y^2)^2}{y^2} \cdot \frac{1}{4s^2} \quad \text{(by our previous inequality).} \blacksquare$$

(3) Computing Some ε_Λ

We now compute ε_Λ for two general classes of Λ. The first is the separated case, where all the λ's are at least 2 away from each other and the second is the unseparated case where each λ is within 2 of the preceding one. It's fairly easy to show that ε_Λ is not affected by changing one or two λ's and thus our two cases include almost all of the general formula cases like 1) $\lambda_k = k^\alpha$ or 2) $\lambda_k = \alpha k$.

The Separated Case.
Here we assume $\Lambda = \{\lambda_0, \lambda_1, \ldots \lambda_N\}$ where $0 = \lambda_0 < \lambda_1 < \cdots < \lambda_N$ and $\lambda_k - \lambda_{k-1} \geq 2$.
We recall that

$$\varepsilon_\Lambda = \max_{\operatorname{Re} z = 1} \left| \frac{1}{z} \prod_{n=1}^{N} \frac{z - \lambda_n}{z + \lambda_n} \right|.$$

Thus

$$\varepsilon_\Lambda^2 = \max_{y} \frac{1}{1 + y^2} \prod_{n=1}^{N} \frac{y^2 + (\lambda_n - 1)^2}{y^2 + (\lambda_n + 1)^2}.$$

Then certainly

$$\varepsilon_\Lambda^2 \geq \prod_{n=1}^{N} \frac{(\lambda_n - 1)^2}{(\lambda_n + 1)^2}$$

which is what we get by setting $y = 0$.

On the other hand,

$$\varepsilon_\Lambda^2 = \max_{y} \frac{(1 - \lambda_1)^2 + y^2}{1 + y^2} \cdot \frac{(1 - \lambda_2)^2 + y^2}{(1 + \lambda_1)^2 + y^2} \cdot \ldots \cdot \frac{1}{(1 + \lambda_N)^2 + y^2}$$

and since

$$\frac{(1 - \lambda_k)^2 + y^2}{(1 + \lambda_{k-1})^2 + y^2} \leq \frac{(1 - \lambda_k)^2}{(1 + \lambda_{k-1})^2}$$

for all y (because $\lambda_k - \lambda_{k-1} \geq 2$) we have

$$\varepsilon_\Lambda^2 \leq \prod_{n=1}^{N} \left(\frac{1 - \lambda_n}{1 + \lambda_n} \right)^2.$$

Thus

$$\varepsilon_\Lambda = \prod_{n=1}^{N} \left| \frac{1 - \lambda_n}{1 + \lambda_n} \right|.$$

Although this is an exact evaluation of ε_Λ, it is sometimes hard to work with. We will now prove the inequalities

$$\frac{1}{2} \exp\left(-2 \sum_{n=1}^{N} \frac{1}{\lambda_n} \right) < \prod_{n=1}^{N} \left(\frac{\lambda_n - 1}{\lambda_n + 1} \right) \leq \exp\left(-2 \sum_{n=1}^{N} \frac{1}{\lambda_n} \right)$$

thereby establishing $\exp(-2 \sum_{n=1}^{N} 1/\lambda_n)$ as the correct approximation index (we recall the "the approximation index" is only determinable to within multiplicative constants).

From the inequality $(1 - x)/(1 + x) \leq e^{-2x}$ for $x \in [0, 1]$ we get

$$\frac{\lambda_k - 1}{\lambda_k + 1} = \frac{1 - \dfrac{1}{\lambda_k}}{1 + \dfrac{1}{\lambda_k}} \leq e^{-2/\lambda_k}$$

and by taking the product of such inequalities we get

$$\prod_{n=1}^{N} \left(\frac{\lambda_n - 1}{\lambda_n + 1}\right) \leq \exp\left(-2 \sum_{n=1}^{N} \frac{1}{\lambda_n}\right).$$

To get the other inequality we proceed as follows. For $x > 0$, $(x - 1)/(x + 1) \exp(2/x)$ is an increasing function of x. Thus, since $\lambda_k \geq 2k$ we have

$$\frac{\lambda_k - 1}{\lambda_k + 1} \exp\left(\frac{2}{\lambda_k}\right) \geq \frac{2k - 1}{2k + 1} \exp\left(\frac{1}{k}\right).$$

Taking the product of these inequalities for $k = 1, 2, \ldots, N$ we get

$$\prod_{k=1}^{N} \left(\frac{\lambda_k - 1}{\lambda_k + 1}\right) \exp\left(2 \sum_{k=1}^{N} \frac{1}{\lambda_k}\right) \geq \frac{1}{2N + 1} \exp\left(\sum_{k=1}^{N} \frac{1}{k}\right)$$

$$\geq \frac{1}{2N + 1} \cdot (N + 1)$$

$$> \frac{1}{2}$$

or

$$\prod_{k=1}^{N} \left(\frac{\lambda_k - 1}{\lambda_k + 1}\right) > \frac{1}{2} \exp\left(-2 \sum_{k=1}^{N} \frac{1}{\lambda_k}\right).$$

One example of such a Λ is $\lambda_k = 2k$; i.e., we choose the even integers. We then get an approximation index of $1/N$, which is the same as with all integers. Thus approximation by even polynomials is as good as approximation by all polynomials. If we were to take $\lambda_k = 3k$ we would get an index of $1/N^{2/3}$ which is more than the index of all polynomials.

Another example of such a Λ is $\lambda_k = k^2$. We then get an approximation index of $e^{2/N}$.

Still another example is given when λ_k is the k-th prime. We then get an index of $1/\log^2 N$.

The Unseparated Case

We now assume that $\Lambda = \{\lambda_0, \lambda_1, \ldots, \lambda_N\}$ with $\lambda_0 = 0$ and $\lambda_k - \lambda_{k-1} < 2$. We will assume that $\sum_{n=1}^{N} \lambda_n > 1/8$ and we will let $S = \sum_{n=1}^{N} \lambda_n$. We recall that

$$\varepsilon_\Lambda^2 = \max_y \frac{1}{1 + y^2} \prod_{n=1}^{N} \frac{y^2 + (\lambda_n - 1)^2}{y^2 + (\lambda_n + 1)^2}$$

and we look at that expression for $y = \sqrt{8S - 1}$. Then

$$\varepsilon_\Lambda^2 \geq \frac{1}{8S} \prod_{n=1}^{N} \frac{8S - 1 + (\lambda_n - 1)^2}{8S - 1 + (\lambda_n + 1)^2}$$

$$= \frac{1}{8S} \prod_{n=1}^{N} \left(1 - \frac{4\lambda_n}{8S - 1 + (\lambda_n + 1)^2}\right)$$

$$\geq \frac{1}{8S} \prod_{n=1}^{N} \left(1 - \frac{4\lambda_n}{8S}\right)$$

$$\geq \frac{1}{8S}\left(1 - \sum_{n=1}^{N} \frac{4\lambda_n}{8S}\right) \left(\text{by } \prod_{i=1}^{M} (1 - a_i) \geq 1 - \sum_{i=1}^{M} a_i \text{ for } a_i \in [0,1]\right)$$

$$= \frac{1}{8S}\left(1 - \frac{1}{2}\right)$$

$$= \frac{1}{16S}.$$

Thus $\varepsilon_\Lambda \geq 1/(4\sqrt{S})$.

We will now prove that $\varepsilon_\Lambda \leq e^{3/2}/(2\sqrt{S})$, thereby establishing

$$\frac{1}{\sqrt{\sum_{n=1}^{N} \lambda_n}}$$

as the approximation index.

To prove the upper bound we divide Λ into 2 sets, Λ_1 and Λ_2, in the following manner. We choose $\lambda_0 \in \Lambda_1$ and then choose as the second element of Λ_1 the largest element of Λ which is not more than 2 greater than λ_0. The third element is chosen similarly, as the largest element of Λ not more than 2 larger than the second element of Λ_1. In this manner we inductively define Λ_1 and Λ_2 is defined as the rest of Λ or $\Lambda - \Lambda_1$. We define $S_1 = \{n : \lambda_n \in \Lambda_1\}$ and $S_2 = \{n : \lambda_n \in \Lambda_2\}$. We let $g_0, g_1, \ldots, g_K$ be the elements of Λ_1 in increasing order. Then we have

(1) $g_K = \lambda_N$.
(2) $g_j - g_{j-1} \leq 2$, $j = 1, \ldots, K$.
(3) $g_j - g_{j-2} > 2$, $j = 2, \ldots, K$.

This last fact follows from the fact that otherwise we would have chosen g_j instead of g_{j-1}. From 3) we easily get

$$g_K = \lambda_N$$
$$g_{K-1} < \lambda_N$$
$$g_{K-2} < g_K - 2 = \lambda_N - 2$$
$$g_{K-3} < g_{K-1} - 2 < \lambda_N - 2$$
$$g_{K-4} < g_{K-2} - 2 < \lambda_N - 4$$

and so on.

We then get $\sum_{n \in S_1} \lambda_n = \sum_{j=0}^{K} g_j \leq 2[\lambda_N + (\lambda_N - 2) + \cdots]$ which certainly is $\leq (\lambda_N + 1)^2$. We also have

$$\prod_{n \in S_1} \frac{y^2 + (\lambda_n - 1)^2}{y^2 + (\lambda_n + 1)^2}$$

$$= \prod_{j=0}^{K} \frac{y^2 + (g_j - 1)^2}{y^2 + (g_j + 1)^2}$$

$$= (y^2 + 1) \cdot \frac{y^2 + (g_1 - 1)^2}{y^2 + 1} \cdot \frac{y^2 + (g_2 - 1)^2}{y^2 + (g_1 + 1)^2} \cdot \cdots \cdot \frac{1}{y^2 + (g_K + 1)^2}$$

$$\leq \frac{y^2 + 1}{y^2 + (g_K + 1)^2} \quad \text{by (2)}$$

$$= \frac{y^2 + 1}{y^2 + (\lambda_N + 1)^2}.$$

On the other hand,

$$\prod_{n \in S_2} \frac{y^2 + (\lambda_n - 1)^2}{y^2 + (\lambda_n + 1)^2} = \prod_{n \in S_2} \left(1 - \frac{4\lambda_n}{y^2 + (\lambda_n + 1)^2}\right)$$

$$\leq \prod_{n \in S_2} \left(1 - \frac{4\lambda_n}{y^2 + (\lambda_N + 1)^2}\right)$$

$$\leq \exp\left(\frac{-4 \sum_{n \in S_2} \lambda_n}{y^2 + (\lambda_N + 1)^2}\right) \quad (\text{by } 1 - u \leq e^{-u})$$

$$= \exp\left(\frac{-4(S - \sum_{n \in S_1} \lambda_n)}{y^2 + (\lambda_N + 1)^2}\right)$$

$$= \exp\left(\frac{-4S}{y^2 + (\lambda_N + 1)^2}\right) \exp\left(\frac{4 \sum_{n \in S_1} \lambda_n}{y^2 + (\lambda_N + 1)^2}\right)$$

$$\leq \exp\left(\frac{-4S}{y^2 + (\lambda_N + 1)^2}\right) \exp\left(\frac{4(\lambda_N + 1)^2}{y^2 + (\lambda_N + 1)^2}\right)$$

$$\leq \exp\left(\frac{-4S}{y^2 + (\lambda_N + 1)^2}\right) e^4.$$

Finally, we have

$$\frac{1}{1 + y^2} \prod_{n=0}^{N} \frac{y^2 + (\lambda_n - 1)^2}{y^2 + (\lambda_n + 1)^2}$$

$$= \frac{1}{1 + y^2} \prod_{n \in S_1} \frac{y^2 + (\lambda_n - 1)^2}{y^2 + (\lambda_n + 1)^2} \prod_{n \in S_2} \frac{y^2 + (\lambda_n - 1)^2}{y^2 + (\lambda_n + 1)^2}$$

$$\leq \frac{1}{1+y^2} \cdot \frac{1+y^2}{y^2+(\lambda_N+1)^2} \cdot \exp\left(\frac{-4S}{y^2+(\lambda_N+1)^2}\right) e^4$$

$$= \frac{e^4}{4S} \cdot \frac{4S}{y^2+(\lambda_N+1)^2} \exp\left(\frac{-4S}{y^2+(\lambda_N+1)^2}\right)$$

$$\leq \frac{e^3}{4S}$$

since $te^{-t} \leq e^{-1}$. Since

$$\varepsilon_\Lambda^2 = \max_y \frac{1}{1+y^2} \prod_{n=0}^{N} \frac{y^2+(\lambda_n-1)^2}{y^2+(\lambda_n+1)^2} \leq \frac{e^3}{4S}$$

we have proven

$$\varepsilon_\Lambda \leq \frac{e^{3/2}}{2\sqrt{S}}$$

and have established

$$\frac{1}{\sqrt{\sum_{n=1}^{N} \lambda_n}}$$

as the approximation index.

Some examples of this kind of Λ are given by $\lambda_k = k^\alpha$, $0 < \alpha \leq 1$. We then get an index of $N^{-((\alpha+1)/2)}$.

If the λ_k satisfy $\lambda_k \leq M$ for all k, we then get an index of $N^{-(1/2)}$.

CHAPTER XII

A UNIFIED TREATMENT OF JACKSON'S THEOREM

We adopt as our setting, some Banach space B of functions defined on $(-\infty, \infty)$ which are periodic with period 2π. We define the translation operator $\cup^t$, for $t \in (-\infty, \infty)$ by $(\cup^t X)(\theta) = X(\theta + t)$.

Definition. A translation space is a Banach space B of functions defined on $(-\infty, \infty)$ such that
(1) $X \in B \Rightarrow X(\theta) = X(\theta + 2\pi)$.
(2) $X \in B \Rightarrow \cup^t X \in B$ for all $t \in (-\infty, \infty)$.
(3) $\cup^t X$ is continuous in the joint variables (t, X); i.e. given $\epsilon > 0$, $X_1 \in B$ and t_1, there is a $\delta > 0$ such that $\|X_1 - X_2\| < \delta$ and

$$|t_1 - t_2| < \delta \Rightarrow \|\cup^{t_1} X_1 - \cup^{t_2} X_2\| < \epsilon.$$

Examples. (1) $C^*[-\pi, \pi]$ (with all functions extended by

$$f(x + 2\pi) = f(x)).$$

(2) $L^P[-\pi, \pi]$, $1 < P < \infty$ (with all functions extended by $f(x + 2\pi) = f(x)$).

$L^\infty[-\pi, \pi]$ is not a translation space since, for $f(x)$ a step function, $\cup^t f$ is not a continuous function of t.

For any t, $\cup^t$ is an operator taking a translation space B into B. We can thus speak of $\|\cup^t\|$ for any t. An important quantity associated with B is defined by $M_B = \sup_t \|\cup^t\|$. Before we prove that this is finite we compute M_B for some translation spaces B. We notice that since

$$\|\cup^t\| = \sup_{X \in B} \frac{\|X(\theta + t)\|}{\|X(\theta)\|}, \quad M_B = \sup_t \sup_{X \in B} \frac{\|X(\theta + t)\|}{\|X(\theta)\|}$$

(1) Let $B = L^P[-\pi, \pi]$, $1 \leq P < \infty$. Then for $f \in L^P[-\pi, \pi]$ and any t, $\|f(\theta + t)\| = \|f(\theta)\|$. Thus $\|\cup^t\| = 1$ and $M_B = 1$.

(2) Let $B = C^*[-\pi, \pi]$ (with sup norm). Then for any $f \in B$, and any t, $\|f(\theta + t)\| = \|f(\theta)\|$. Thus $\|\cup^t\| = 1$ and $M_B = 1$.

(3) Let $B = C^*[-\pi, \pi]$ with $\|f\| = \max_{|\theta| \leq \pi}(1 + |\theta|)|f(\theta)|$.

Obviously $\|f(t + \theta)\| \leq (1 + \pi)\|f(\theta)\|$. However, if we take a function $f(\theta)$ which has its maximum at 0 and falls away rapidly to 0, then $\|f\|$ is equal to $|f(0)|$ while $\|\cup^\pi f\| = \max_{|\theta| \leq \pi}(1 + |\theta|)|f(\theta + \pi)|$ which is about equal to $(1 + \pi)|f(0)|$. Thus $M_B = 1 + \pi$.

LEMMA 1. *In any translation space B, $M_B = \sup_t \|\cup^t\|$ is finite.*

Proof. Assume the lemma is false. Then, we can find $\{t_n\}$, such that $\|\cup^{t_n}\| > n$. Therefore, there is $X_n \in B$ such that $\|\cup^{t_n} X_n\| > n \|X_n\|$, $n = 1, 2, \cdots$. Without loss of generality, we can assume $|t_n| \leq \pi$, and then, because of compactness, we can extract a convergent subsequence, which we might as well assume is $\{t_n\}$ itself, and t_n converges to τ.

Let $Y_n = X_n/(n\|X_n\|)$. Then $\|Y_n\| = 1/n$ and $\|\cup^{t_n} Y_n\| > 1$. Given $\varepsilon = 1/2$, find the corresponding δ for $(\tau, 0)$;. i.e. $\|z - 0\| < \delta$ and $\|t - \tau\| < \delta \Rightarrow \|\cup^t z - \cup^\tau 0\| < 1/2$, or $\|z\| < \delta$ and $\|t - \tau\| < \delta \Rightarrow \|\cup^t z\| < 1/2$.

By taking a large n, we can have $\|Y_n\| = 1/n < \delta$ and $|t_n - \tau| < \delta$. However, $\|\cup^{t_n} Y_n\| > 1$, contradicting the continuity of $\cup^t X$. ∎

In this abstract setting we have one analog of a theorem by Bernstein.

THEOREM 2. *Let $P(\theta)$ be an n-th degree trigonometric polynomial which lies in B. Then $P'(\theta) \in B$ and $\|P'(\theta)\| \leq n M_B \|P\|$.*

Proof: Let $\mathscr{L}$ be an element of B^*, the dual space of B. We form $\varphi(t) = \mathscr{L}(\cup^t P)$. We observe that $\varphi(t)$ is a trigonometric polynomial and $\sup_t |\varphi(t)| \leq \|\mathscr{L}\| M_B \|P\|$. Let ε be given > 0 and consider $(\varphi(\varepsilon) - \varphi(0))/\varepsilon$. On the one hand, it is equal to $\varphi'(\zeta)$, $0 < \zeta < \varepsilon$, and on the other, it is equal to $\mathscr{L}((\cup^\varepsilon P - P)/\varepsilon)$. Thus

$$\left|\mathscr{L}\left(\frac{\cup^\varepsilon - P}{\varepsilon}\right)\right| = |\varphi'(\zeta)| \leq n \sup_t |\varphi(t)| \leq \|\mathscr{L}\| n M_B \|P\|$$

Since this holds for any $\mathscr{L} \in B^*$, we can conclude that

$$\left\|\frac{\cup^\varepsilon P - P}{\varepsilon}\right\| \leq n M_B \|P\|.$$

However as $\varepsilon \to 0$,

$$\left(\frac{\cup^\varepsilon P - P}{\varepsilon}\right)(\theta) = \frac{P(\theta + \varepsilon) - P(\varepsilon)}{\varepsilon}$$

converges to $P'(\theta)$ uniformly. Since, on a finite dimensional vector space, all norms are equivalent, we get that $P'(\theta) \in B$ and $(\cup^\varepsilon P - P)/\varepsilon$ converges to P' in B. Combining this with the previous inequality for $(\cup^\varepsilon P - P)/\varepsilon$, we get $\|P'\| \leq n M_B \|P\|$. ∎

For a fixed $X \in B$, the expression $\cup^t X$ may be thought of as a B-valued continuous function of the variable t. As such it is susceptible to Riemann integration. Indeed, for any periodic measure du on $[-\pi, \pi]$, the integral $\int \cup^t X \, du(t)$ makes perfect sense as an element of B. In particular, we may speak of $\int K(t) \cup^t X \, dt$ where $K(t)$ is a continuous periodic function. One further remark will be borrowed from Riemann integral theory and this is the inequality

$$\left\| \int \cup^t X \, du(t) \right\| \leq \int \| \cup^t X \| \, |du(t)|,$$

which is easily proven by considering Riemann sums.

LEMMA 3. *Let k be any integer and let $X \in B$. Then there exists a C such that*

$$\int_{-\pi}^{\pi} e^{-ikt} \cup^t X \, dt$$

(considered as a function of θ) $= C e^{ik\theta}$

Proof. Call

$$\int_{-\pi}^{\pi} e^{-ikt} \cup^t X \, dt = Y$$

and apply $\cup^s$. By the boundedness of this operator, we obtain

$$\cup^s Y = \int_{-\pi}^{\pi} e^{-ikt} \cup^{t+s} X \, dt$$

$$= e^{iks} \int_{-\pi}^{\pi} e^{-ik(t+s)} \cup^{(t+s)} X \, dt$$

$$= e^{iks} \int_{-\pi+s}^{\pi+s} e^{-ikt} \cup^t X \, dt$$

$$= e^{iks} \int_{-\pi}^{\pi} e^{-ikt} \cup^t X \, dt$$

$$= e^{iks} Y.$$

Thus $Y(\theta + s) = e^{iks} Y(\theta)$. Now set $\theta = 0$, and we get the desired result. ∎

Corollary. If $K(t) \in \mathscr{T}_n$, then $\int_{-\pi}^{\pi} K(t) \cup^t X \, dt \in \mathscr{T}_n$.

Definition. The modulus of continuity of an element $X \in B$ is defined,

for $\delta > 0$, by $\omega(X, \delta) = \sup_{|t| \leq \delta} \|\bigcup^t X - X\|$.
$\omega(X, \delta)$ has the following properties:

(1) $\omega(X, \delta)$ is non-decreasing.

(2) $\omega(X, 0^+) = 0$.

(3) $\omega(X, \delta_1 + \delta_2 + \ldots + \delta_n) \leq M_B \sum_{i=1}^{n} \omega(X, \delta_i)$.

(3) is the only one that requires proof.

$$\bigcup^{\delta_1 + \cdots + \delta_n} X - X = \bigcup^{\delta_1 + \cdots + \delta_n} X - \bigcup^{\delta_2 + \delta_3 + \cdots \delta_n} X$$
$$+ \bigcup^{\delta_2 + \cdots + \delta_n} X - \bigcup^{\delta_3 + \cdots + \delta_n} X$$
$$+ \cdots +$$
$$+ \bigcup^{\delta_n} X - X.$$

Using the triangle inequality, we get

$$\|\bigcup^{\delta_1 + \cdots + \delta_n} X - X\| \leq \|\bigcup^{\delta_2 + \cdots + \delta_n}(\bigcup^{\delta_1} X - X)\|$$
$$+ \|\bigcup^{\delta_3 + \cdots + \delta_n}(\bigcup^{\delta_2} X - X)\|$$
$$+ \cdots +$$
$$+ \|\bigcup^{\delta_n} X - X\|$$
$$\leq M_B \|\bigcup^{\delta_1} X - X\| + M_B \|\bigcup^{\delta_2} X - X\|$$
$$+ \cdots + M_B \|\bigcup^{\delta_n} X - X\|$$
$$\leq M_B \sum_{i=1}^{n} \omega(X, \delta_i) \quad \blacksquare$$

Corollary. $\omega(X, |t|) \leq M_B \omega(X, 1/n)(n|t| + 1)$.

Proof. $\omega(X, |t|) = \omega(X, (1/n)n|t|) \leq \omega(X, (1/n)([n|t|] + 1))$
$$\leq M_B([n|t|] + 1)\omega(X, (1/n))$$
$$\leq M_B(n|t| + 1)\omega(X, (1/n)) \quad \blacksquare$$

We are now ready to prove a form of:

JACKSON'S THEOREM 4. Let $X \in B$. Then, there exists

$$P \in \mathscr{T}_n \cap B,$$

such that

$$\|X - P\| \leq 17 \, M_B \, \omega(X, (1/n)).$$

Proof. Let $K_n(\theta)$ be the Jackson kernel. The following properties of $K_n(\theta)$ were proven in Chapter IV:

(1) $K_n(\theta) \in \mathscr{T}_n$.

(2) $K_n(\theta) \geq 0$.

(3) $\int_{-\pi}^{\pi} K_n(\theta)\, d\theta = 1$.

(4) $\int_{-\pi}^{\pi} |\theta| K_n(\theta)\, d\theta \leq 16/n$.

Let $P = \int_{-\pi}^{\pi} K_n(t) \cup^t X\, dt$. By property (1) of $K_n(t)$ and the corollary to lemma 3, we know that $P \in \mathscr{T}_n \cap B$. By property (3), we have that

$$X - P = \int_{-\pi}^{\pi} K_n(t)\, (X - \cup^t X)\, dt\,.$$

Therefore,

$$\begin{aligned}\|X - P\| &\leq \int_{-\pi}^{\pi} K_n(t) \|X - \cup^t X\|\, dt \\ &\leq \int_{-\pi}^{\pi} K_n(t)\, \omega(X, |t|)\, dt \\ &\leq M_B\, \omega(X, (1/n)) \int_{-\pi}^{\pi} K_n(t)\, (n|t| + 1)\, dt \\ &\leq M_B\, \omega(X, (1/n))\, 17 \quad\blacksquare\end{aligned}$$

BIBLIOGRAPHICAL NOTES

Two books on approximation theory, whose union contains many interesting and important topics in approximation theory, are [6] and [19]. In particular, [6] has an extremely extensive bibliographic section including many historic notes.

Chapter I
Weierstrass' theorem was first published in [36].
The well known theorem about the convergence of Cesàro sums, as well as many other results in Fourier series, can be found in [3] and [38].
Lebesgue's proof of Weirstrass' theorem was originally published in [18].
More information about the Bernstein polynomials can be found in many places, two of which are [20] and the first chapter of [19].
Both [6] and [32] discuss some aspects and properties of the Tchebychev polynomials.

Chapter IV
Some forms of Jackson's theorems were originally proven in [16]. Nowadays, virtually every book on approximation theory has some reference to them.
An extensive discussion of Gaussian quadrature (also known as mechanical quadrature), as well as a development of Legendre polynomials is found in [32].
Favard's theorem was originally published in [9].

Chapter V
Bernstein's theorems can be found in [4].
The development of the Zygmund class was originally formulated in [37].

Chapter VI
The operators D_n were originally developed by de La Vallée Poussin in [7]. A further discussion of these operators and the lower bound of approximation for $E_n^*(|\sin x|)$ is in [17].
The Uniform Boundedness Theorem (also known as the Banach-Steinhaus theorem) is discussed in [1].
The minimality of the Fourier projection is also known as the Lozinski-Kharshiladze theorem, the Kharshiladze-Lozinski theorem, and as the Lozinski theorem (see [6] and [17]). It is based on the representation of the Fourier sums as an integral involving U^α as in lemma 6.6. This representation was first

done by Marcinkiewicz in [22].
Korovkin's theorems can be found in Korovkin [17].
The papers of Shisha and Mond are in [30] and [31].

Chapter VII
The upper and lower estimates for $R_n(|x|)$ are contained in [26]. Many papers followed this one and considered approximating classes of functions of which $|x|$ is a member. Examples of this are [5], [12], [13], [15], [33], [34], [35].

Chapter VIII
The last third of this chapter (from theorem 9 on) is contained in [10]; [11] is a related paper.

Chapter IX
This chapter is essentially in [27].

Chapter X
This form of the Riesz representation theorem can be found in [29].
Müntz's theorem was originally proven in [23].
One source for the Paley-Wiener theorem is [28].

Chapter XI
The first paper on a Müntz-Jackson theorem was [25], and it dealt with L^2 approximation when $\lambda_{K+1} - \lambda_K \geq 2$ for all K. Following that paper, [2], [14], and [21] (among others) were written, improving on [25]. The Müntz-Jackson theorem in this chapter, which is the " ultimate " one, is essentially [24].

BIBLIOGRAPHY

[1] G. Bachman and L. Narici: *Functional Analysis*, Academic Press, New York: 1966.
[2] J. Bak and D. J. Newman: *Müntz-Jackson Theorems in $L^p[0, 1]$ and $C[0,1]$, Am. J. Mathematics*, 94 (1972), 437-457.
[3] N. K. Bary: *A Treatise on Trigonometric Series*, Pergamon Press, New York: 1964.
[4] S. N. Bernstein: Collected Works (in Russian), *Akad. Nauk SSSR*, *1*: 1952, *2*: 1954. (Volume 1 has been translated into English and published in 1958 by the Atomic Energy Commission translations.)
[5] A. P. Bulanov: On the Order of Approximation of Convex Functions by Rational Functions, *Math USSR Izvestija*, 3 (1969), 1067-1080.
[6] E. W. Cheney: *Introduction to Approximation Theory*, McGraw-Hill, New York: 1966.
[7] C. J. de la Vallée Poussin: Sur la meilleure approximation des fonctions . . . . , *Compt. Rend.*, 166 (1918), 799-802.
[8] A. Erdelyi, et al.: *Higher Transcendental Functions*, Vol. 2, McGraw Hill, New York: 1955.
[9] J. Favard, Sur les meilleures procédés d'approximation de certaines classes . . . , *Bull. Sci. Math. France*, 61 (1937), 209-224, 243-256.
[10] R. Feinerman: A best two-dimensional space of approximating functions, *J. Approximation Theory*, 3 (1970), 50-58.
[11] R. Feinerman: A best two-dimensional space of approximating functions, II, *J. Approximation Theory*, 4 (1971), 328-331.
[12] G. Freud and J. Szabados: On rational approximation, *Stud. Scient. Math. Hung.*, 2 (1967), 215-219.
[13] G. Freud: A remark concerning the rational approximation to $|x|$, *Stud. Scient. Math. Hung.*, 2 (1967), 115-117.
[14] T. Ganelius and S. Westlund: *The Degree of Approximation in Müntz's Theorem*, Proceedings of the International Conference on Mathematical Analysis, Jyvaskyla, Finland, 1970.
[15] A. A. Goncar: On the rapidity of rational approximation of continuous functions with characteristic singularities, *Math. USSR-Sbornik*, 2 (1967), 561-568.
[16] D. Jackson: *The Theory of Approximation*, A. M. S. Colloquium Publications, 1930.
[17] P. P. Korovkin: *Linear Operators and Approximation Theory*, Hindustan Pub. Corp. 1960.

[18] H. Lebesgue: Sur l'approximation des fonctions, *Bull. Soc. Math. France,* 22 (1898), 278-287.
[19] G. G. Lorentz: *Approximation of Functions,* Holt, Rinehart and Winston, New York: 1966.
[20] G. G. Lorentz: *Bernstein Polynomials,* University of Toronto Press, Toronto: 1955.
[21] K. Lung: *On Müntz-Jackson Theorems,* Doctoral thesis at SUNY, Stony Brook, 1971.
[22] J. Marcinkiewicz: Quelques remarques sur l'interpolation, *Acta Litt. Scient., Szeged 8* (1937), 127-130.
[23] C. Müntz: *Uber Den Approximationssatz Von Weierstrass,* Schwarz-Festschrift, (1914), 302-312.
[24] D. J. Newman: A general Müntz-Jackson Theorem, *Am. J. Mathematics,* to be published.
[25] D. J. Newman: A Müntz-Jackson *Am. J. Mathematics,* 87 (1965), 940-944.
[26] D. J. Newman: Rational Approximation to $|x|$, *Mich. Math. J.,* 11 (1964), 11-14.
[27] D. J. Newman and H. S. Shapiro: *Jackson's Theorem in Higher Dimensions,* in " On Approximation Theory, Proceedings of the Conference at Oberwolfach, 1963, Basel " Birkhauser, 1964.
[28] R. Paley and N. Wiener: *Fourier Transforms in the Complex Domain,* A.M.S. Colloquium Publications, Vol. XIX, 1934.
[29] F. Riesz and B. Sz. Nagy: *Functional Analysis,* Ungar Publishing Co. New York: 1955.
[30] O. Shisha and B. Mond: The degree of approximation to periodic functions by linear positive operators, *J. Approximation Theory,* 1 (1968), 335-339.
[31] O. Shisha and B. Mond: The degree of convergence of sequences of linear positive operators, *Proc. Natl. Acad. Sci.,* 60 (1968), 1196-1200.
[32] G. Szego: *Orthogonal Polynomials,* A.M.S. Colloquium Publications, Ed. 2, 1959.
[33] P. Szüsz and P. Turán: On the constructive theory of functions I, *Publ. Math. Inst. Hung. Akad. Sci.,* 9 (1964), 495-502.
[34] P. Szüsz and P. Turán: On the constructive theory of functions II, *Stud. Scient. Math. Hung.,* 1 (1966), 65-69.
[35] P. Szüsz and P. Turán: On the constructive theory of functions III, *Stud. Scient. Math. Hung.,* 1 (1966), 315-322.
[36] K. Weierstrass: *Uber die Analytische Darstellbarkeit Sogennanter Willkurlieher Funktionen einer Reelen Veraderlichen,* Berliner Berichte, (1885), 633-639, 789-805.
[37] A. Zygmund: Smooth functions, *Duke Math. J.,* 12 (1945), 47-76.
[38] A. Zygmund: *Trigonometric Series,* Ed. 2, Cambridge University Press, New York: (1959).

INDEX OF SYMBOLS

B^*, 139
B_k, 97
$B_n(f, x)$, 6
$B(z)$, 122
$c(\Lambda)$, 123
$C[a, b]$, 1
$C_{\omega(\delta)}$, 15
$C_{\omega(\delta), M}$, 15
$C^*[-\pi, \pi]$, 2
$C^p[0, 1]$, 39
$C^{p*}[-\pi, \pi]$, 44
$C(T)$, 98
$d\sigma$, 105
$d(T)$, 90
$D_N(f, x)$, 58
$E_g(\mathscr{S})$, 91
$E_\Lambda(f)$, 121
$E_n(C)$, 16
$E_n^*(C)$, 34
$E_n(f)$, 16
$E_n^*(f)$, 33
$E_n(\mathscr{S}(T))$, 98
$E_\phi(f)$, 81
$E_\phi(\mathscr{S})$, 83
ϵ_Λ, 121, 122
ϵ_n, 85
$\epsilon_n(M)$, 85
$\|f\|$, 2
$F_n(x)$, 46
$J_n(f, x)$, 41
$K_n(x)$, 41
$L_N((l - x)^2, x)$, 68

M_B, 138
n_k, 99
n_k^*, 99
$\omega(f, \delta)$, 13, 87
$\omega(X, \delta)$, 141
$\mathcal{P}_n$, 16
$\mathcal{P}_n^k$, 98
R^k, 97
$R_n(C)$, 71
$R_n(f)$, 71
$\mathcal{R}_n$, 71
ρ_Λ, 121
$\rho_y(x)$, 103
$\rho(x, T)$, 86
S_{k-1}, 97
$S_n(f, x)$, 40
$\|S_N\|$, 61
$\mathscr{S}$, 13, 83
$\mathscr{S}^*$, 34
$\mathscr{S}(M)$, 83
$\mathscr{S}_0(M)$, 91
$\sigma_n(f, x)$, 41
$\text{sp}\{\varphi_1, \varphi_2, \cdots, \varphi_n\}$, 81
T^c, 90
$T_n(x)$, 2
$T(x)$, 90
$\mathfrak{J}_n$, 33
W_k, 97
$x \cdot y$, 97
Z, 53
$[\ ]$, 41
U^α, 65
$\|\|\|_1$, 81

INDEX

Approximation index, 121, 134, 135, 137
Ascoli-Arzela theorem, 15

Bernstein, 49, 51
Bernstein polynomials, 6–11, 13, 16, 70

Carlson's lemma, 129
Cesàro sums, 1, 2, 41, 57–60
Codimension, 117–120
Compact, 15, 24, 25
Completeness, 108–120

Descartes rule of signs, 21
Dini-Lipschitz theorem, 64
Dirichlet kernel, 40

$E_0(\mathscr{S})$, 17, 26
$E_1(\mathscr{S})$, 17, 32, 33
$E_n(\mathscr{S})$, 17, 33, 35, 36, 38, 72
$E_n^*(\mathscr{S}^*)$, 33–35, 44, 46, 48, 57
Equicontinuous, 15
Equivalent norms, 22

Favard's theorem, 46–48
Fejér kernel, 41, 42
Fourier series, 1, 57, 59, 61, 64, 65

Gaussian quadrature, 35, 36

Hahn-Banach theorem, 109, 125

Jackson kernel, 41–43, 46, 53, 54, 124, 141, 142
Jackson operator, 41, 42
Jackson's theorems, 33, 35–49, 51, 70, 72, 84, 97–107, 124, 138–142

Korovkin's theorem, 68–70

Lebesgue, 4–6
Legendre polynomials, 35, 37
Lip α, 38, 49, 51, 52, 55, 56, 65, 72, 73, 75

Minkowski's inequality, 82
Modulus of continuity, 13–15
Mond, 68
Müntz-Jackson theorem, 121–137
Müntz's theorem, 108, 111–115

Orthant, 83, 86, 87

Paley Wiener theorem, 114, 128
Polynomials, algebraic, 16
    derivatives of, 20, 21, 75
    of best approximation, 24–33, 67
    Bernstein, 6–11, 13, 16, 70
    Legendre, 35, 37
    rational, 71–80
    Tchebychev, 2–4, 20, 21, 31, 104
    trigonometric, 33
        derivatives of, 18
        of best approximation, 33, 34, 66, 67
    zeroes of, 17
Pre-compact, 15, 16

Rational polynomials, 71–80
Riesz representation theorem, 109, 110

Shisha, 68
Strictly convex, 82, 83

Tchebychev polynomials, 2–4, 20, 21, 31, 104

Uniform boundedness theorem, 64, 66

Weierstrass theorem, 1, 2, 4, 7, 11, 16, 108

Zygmund, 53, 54

QA
221
F44.

MAR 5 1975